6000 Practice Problems

100 Days of Timed Tests

HORIZONTAL MULTIPLICATION & DIVISION

Facts 1-12

Ages 7-10

109 Pages

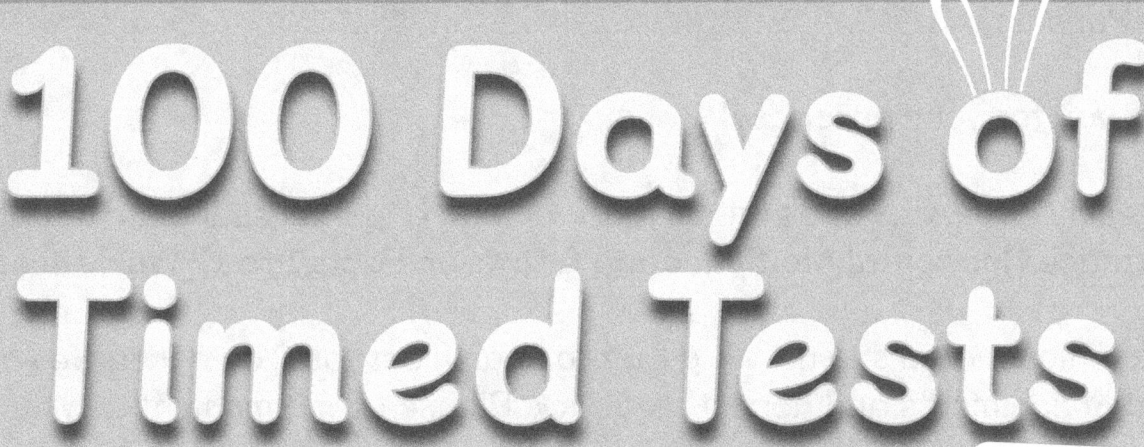

Grade 3-5

MATH DRILLS   SPEED MATH PRACTICE

I0616448

Dear Parents,

Thank you for your purchase!

I sincerely hope, this book will be more helpful and interesting for the kids to practice **Horizontal Multiplication & Division Facts 1 to 12 Worksheets.**

Your opinion matters to us. We'd love to hear from you! You can leave your valuable comments and honest feedback. Please take a moment to write a review. We appreciate your support.

Email us at abczbook@gmail.com with the title **"Horizontal Multiplication & Division Facts 1 to 12 Workbook"** and get your free practice worksheets!

Hopefully, your kids will enjoy this book.

Enjoy Learning!

abcZbook Press
www.abczbook.com

abcZbook Press
www.abczbook.com

This book belongs to

_____

Grade: _____

School:_____

A few minutes of math workout every day will help the children master the math skills.

This **"Horizontal Multiplication & Division Facts 1 to 12"** is the beginner level math practice workbook for Grade 3-5. This is suitable for Ages 7-10.

This practice workbook contains a set of multiplication and division problems written horizontally (i.e left to right) up to number 12. This book is designed to help kids to improve multiplication and division facts up to 12.

These sets of math horizontal practice worksheets are designed to test multiplication without regrouping and division without remainder. The kids can challenge themselves with the timed test problems. This book mainly focuses on improving speed, problem solving skills, memory power, and building confidence levels.

This book also contains **Answer Key sheets** at the end of the book, so that you can quickly check the kid's answer. In this book, there are 60 problems to be solved daily and a total of 100 pages of Timed test practice worksheets. It helps the kids to perform consistently and trained to be excellent in multiplication and division problems.

Get more practice from abcZbook's Addition, Subtraction, Multiplication and Division Timed Tests workbooks.

Table of Contents

No.	Sections
1	Problems: Horizontal Multiplication Facts 1 to 12
2	Problems: Horizontal Division Facts 1 to 12
3	Multiplication Answer Key Sheet
4	Division Answer Key Sheet
5	Multiplication Tables 1 to 12
6	Division Tables 1 to 12
7	Certificate of Excellence

(1) 7 x 2 =

(2) 8 x 3 =

(3) 10 x 2 =

(4) 4 x 5 =

(5) 5 x 3 =

(6) 4 x 4 =

(7) 8 x 4 =

(8) 4 x 3 =

(9) 1 x 11 =

(10) 10 x 7 =

(11) 12 x 3 =

(12) 6 x 12 =

(13) 9 x 3 =

(14) 3 x 10 =

(15) 4 x 4 =

(16) 6 x 7 =

(17) 11 x 1 =

(18) 1 x 8 =

(19) 4 x 6 =

(20) 8 x 5 =

(21) 2 x 1 =

(22) 1 x 5 =

(23) 6 x 5 =

(24) 4 x 1 =

(25) 1 x 7 =

(26) 7 x 12 =

(27) 10 x 5 =

(28) 1 x 3 =

(29) 6 x 4 =

(30) 3 x 2 =

(31) 12 x 11 =

(32) 9 x 8 =

(33) 8 x 10 =

(34) 3 x 1 =

(35) 6 x 1 =

(36) 11 x 8 =

(37) 9 x 2 =

(38) 11 x 12 =

(39) 5 x 8 =

(40) 9 x 10 =

(41) 12 x 8 =

(42) 1 x 6 =

(43) 4 x 11 =

(44) 9 x 11 =

(45) 10 x 6 =

(46) 5 x 6 =

(47) 10 x 2 =

(48) 4 x 9 =

(49) 8 x 1 =

(50) 2 x 10 =

(51) 1 x 8 =

(52) 3 x 4 =

(53) 2 x 9 =

(54) 5 x 5 =

(55) 11 x 7 =

(56) 11 x 10 =

(57) 5 x 2 =

(58) 12 x 6 =

(59) 11 x 3 =

(60) 7 x 12 =

(1) 11 x 11 =
(2) 5 x 3 =
(3) 12 x 9 =

(4) 11 x 1 =
(5) 9 x 5 =
(6) 2 x 5 =

(7) 3 x 2 =
(8) 2 x 3 =
(9) 2 x 6 =

(10) 4 x 7 =
(11) 12 x 9 =
(12) 8 x 11 =

(13) 8 x 3 =
(14) 7 x 8 =
(15) 3 x 1 =

(16) 1 x 3 =
(17) 9 x 10 =
(18) 6 x 9 =

(19) 6 x 1 =
(20) 5 x 11 =
(21) 2 x 10 =

(22) 1 x 10 =
(23) 8 x 6 =
(24) 8 x 1 =

(25) 12 x 12 =
(26) 5 x 3 =
(27) 6 x 11 =

(28) 6 x 11 =
(29) 11 x 2 =
(30) 1 x 10 =

(31) 12 x 6 =
(32) 9 x 7 =
(33) 4 x 12 =

(34) 8 x 2 =
(35) 10 x 10 =
(36) 4 x 1 =

(37) 11 x 10 =
(38) 11 x 1 =
(39) 7 x 6 =

(40) 6 x 4 =
(41) 3 x 1 =
(42) 11 x 8 =

(43) 5 x 8 =
(44) 11 x 10 =
(45) 7 x 2 =

(46) 1 x 9 =
(47) 5 x 4 =
(48) 2 x 1 =

(49) 5 x 10 =
(50) 7 x 10 =
(51) 3 x 3 =

(52) 5 x 11 =
(53) 2 x 3 =
(54) 6 x 5 =

(55) 10 x 12 =
(56) 1 x 9 =
(57) 12 x 12 =

(58) 4 x 2 =
(59) 10 x 11 =
(60) 1 x 6 =

(1) 10 x 10 = (2) 1 x 7 = (3) 1 x 9 =

(4) 11 x 7 = (5) 2 x 9 = (6) 8 x 3 =

(7) 4 x 9 = (8) 5 x 5 = (9) 5 x 7 =

(10) 11 x 3 = (11) 3 x 3 = (12) 4 x 1 =

(13) 2 x 8 = (14) 3 x 11 = (15) 8 x 4 =

(16) 5 x 6 = (17) 4 x 8 = (18) 5 x 8 =

(19) 12 x 2 = (20) 5 x 12 = (21) 1 x 9 =

(22) 10 x 8 = (23) 11 x 10 = (24) 3 x 6 =

(25) 1 x 1 = (26) 4 x 9 = (27) 6 x 7 =

(28) 5 x 7 = (29) 3 x 12 = (30) 12 x 11 =

(31) 9 x 9 = (32) 9 x 1 = (33) 9 x 8 =

(34) 10 x 1 = (35) 11 x 11 = (36) 9 x 6 =

(37) 3 x 2 = (38) 2 x 6 = (39) 9 x 3 =

(40) 8 x 4 = (41) 11 x 12 = (42) 1 x 12 =

(43) 2 x 8 = (44) 7 x 11 = (45) 7 x 10 =

(46) 4 x 10 = (47) 7 x 3 = (48) 7 x 5 =

(49) 10 x 4 = (50) 3 x 3 = (51) 9 x 5 =

(52) 3 x 4 = (53) 3 x 12 = (54) 7 x 8 =

(55) 9 x 2 = (56) 6 x 7 = (57) 12 x 11 =

(58) 7 x 11 = (59) 2 x 12 = (60) 12 x 11 =

Day:	4	Date:
Name:		Time: :
		Score: /60
		Rating: ☆☆☆☆☆☆

(1) 6 x 9 =

(2) 3 x 2 =

(3) 4 x 2 =

(4) 3 x 12 =

(5) 11 x 10 =

(6) 5 x 5 =

(7) 10 x 4 =

(8) 2 x 12 =

(9) 2 x 8 =

(10) 3 x 7 =

(11) 9 x 3 =

(12) 10 x 5 =

(13) 2 x 12 =

(14) 12 x 11 =

(15) 7 x 8 =

(16) 6 x 6 =

(17) 9 x 6 =

(18) 10 x 2 =

(19) 4 x 2 =

(20) 8 x 1 =

(21) 1 x 5 =

(22) 1 x 10 =

(23) 8 x 5 =

(24) 8 x 1 =

(25) 10 x 8 =

(26) 3 x 9 =

(27) 5 x 1 =

(28) 11 x 4 =

(29) 1 x 4 =

(30) 9 x 4 =

(31) 11 x 1 =

(32) 9 x 2 =

(33) 2 x 10 =

(34) 4 x 1 =

(35) 4 x 12 =

(36) 11 x 10 =

(37) 11 x 5 =

(38) 11 x 10 =

(39) 5 x 2 =

(40) 9 x 3 =

(41) 5 x 5 =

(42) 4 x 3 =

(43) 5 x 12 =

(44) 6 x 5 =

(45) 1 x 8 =

(46) 11 x 8 =

(47) 7 x 6 =

(48) 11 x 11 =

(49) 2 x 1 =

(50) 7 x 9 =

(51) 11 x 8 =

(52) 12 x 11 =

(53) 5 x 10 =

(54) 2 x 2 =

(55) 2 x 8 =

(56) 5 x 3 =

(57) 7 x 8 =

(58) 3 x 1 =

(59) 8 x 6 =

(60) 5 x 10 =

(1) $3 \times 6 =$

(2) $1 \times 8 =$

(3) $7 \times 6 =$

(4) $8 \times 5 =$

(5) $11 \times 10 =$

(6) $3 \times 2 =$

(7) $6 \times 2 =$

(8) $9 \times 2 =$

(9) $4 \times 2 =$

(10) $11 \times 10 =$

(11) $7 \times 9 =$

(12) $8 \times 9 =$

(13) $2 \times 2 =$

(14) $9 \times 5 =$

(15) $6 \times 9 =$

(16) $11 \times 4 =$

(17) $12 \times 9 =$

(18) $3 \times 1 =$

(19) $8 \times 3 =$

(20) $7 \times 12 =$

(21) $7 \times 8 =$

(22) $6 \times 10 =$

(23) $1 \times 5 =$

(24) $9 \times 11 =$

(25) $4 \times 3 =$

(26) $6 \times 8 =$

(27) $7 \times 1 =$

(28) $7 \times 12 =$

(29) $1 \times 10 =$

(30) $9 \times 9 =$

(31) $8 \times 7 =$

(32) $9 \times 5 =$

(33) $10 \times 12 =$

(34) $3 \times 9 =$

(35) $5 \times 1 =$

(36) $3 \times 5 =$

(37) $8 \times 3 =$

(38) $12 \times 11 =$

(39) $2 \times 6 =$

(40) $8 \times 2 =$

(41) $4 \times 2 =$

(42) $10 \times 10 =$

(43) $5 \times 8 =$

(44) $7 \times 5 =$

(45) $11 \times 2 =$

(46) $4 \times 11 =$

(47) $5 \times 5 =$

(48) $1 \times 3 =$

(49) $8 \times 6 =$

(50) $4 \times 4 =$

(51) $6 \times 10 =$

(52) $1 \times 3 =$

(53) $3 \times 10 =$

(54) $10 \times 12 =$

(55) $10 \times 3 =$

(56) $12 \times 6 =$

(57) $7 \times 11 =$

(58) $10 \times 10 =$

(59) $10 \times 3 =$

(60) $12 \times 3 =$

(1) 11 x 2 =   (2) 5 x 6 =   (3) 7 x 9 =

(4) 10 x 12 =   (5) 10 x 10 =   (6) 10 x 7 =

(7) 2 x 3 =   (8) 3 x 11 =   (9) 3 x 3 =

(10) 8 x 7 =   (11) 6 x 4 =   (12) 7 x 1 =

(13) 2 x 10 =   (14) 8 x 12 =   (15) 12 x 8 =

(16) 9 x 8 =   (17) 6 x 9 =   (18) 11 x 6 =

(19) 5 x 5 =   (20) 1 x 9 =   (21) 11 x 1 =

(22) 5 x 9 =   (23) 12 x 12 =   (24) 4 x 3 =

(25) 3 x 7 =   (26) 1 x 10 =   (27) 4 x 6 =

(28) 12 x 4 =   (29) 6 x 1 =   (30) 7 x 10 =

(31) 9 x 12 =   (32) 3 x 8 =   (33) 11 x 3 =

(34) 1 x 7 =   (35) 1 x 4 =   (36) 1 x 2 =

(37) 7 x 12 =   (38) 7 x 4 =   (39) 4 x 10 =

(40) 9 x 6 =   (41) 11 x 11 =   (42) 10 x 7 =

(43) 1 x 7 =   (44) 7 x 9 =   (45) 2 x 12 =

(46) 7 x 8 =   (47) 10 x 12 =   (48) 6 x 5 =

(49) 1 x 11 =   (50) 4 x 8 =   (51) 7 x 6 =

(52) 3 x 9 =   (53) 2 x 2 =   (54) 9 x 9 =

(55) 12 x 5 =   (56) 10 x 1 =   (57) 4 x 10 =

(58) 8 x 1 =   (59) 8 x 4 =   (60) 10 x 3 =

(1) 6 x 12 =   (2) 9 x 6 =   (3) 9 x 8 =

(4) 5 x 1 =   (5) 7 x 11 =   (6) 2 x 12 =

(7) 10 x 12 =   (8) 5 x 4 =   (9) 2 x 9 =

(10) 3 x 8 =   (11) 7 x 2 =   (12) 1 x 3 =

(13) 8 x 5 =   (14) 12 x 10 =   (15) 10 x 4 =

(16) 8 x 3 =   (17) 8 x 1 =   (18) 8 x 3 =

(19) 3 x 12 =   (20) 6 x 9 =   (21) 10 x 6 =

(22) 10 x 8 =   (23) 4 x 6 =   (24) 12 x 7 =

(25) 11 x 4 =   (26) 11 x 7 =   (27) 2 x 5 =

(28) 6 x 3 =   (29) 5 x 4 =   (30) 10 x 5 =

(31) 2 x 5 =   (32) 11 x 8 =   (33) 7 x 2 =

(34) 8 x 7 =   (35) 7 x 2 =   (36) 5 x 2 =

(37) 5 x 4 =   (38) 8 x 2 =   (39) 11 x 11 =

(40) 7 x 10 =   (41) 9 x 12 =   (42) 12 x 12 =

(43) 2 x 9 =   (44) 7 x 6 =   (45) 4 x 3 =

(46) 3 x 2 =   (47) 1 x 5 =   (48) 12 x 1 =

(49) 2 x 12 =   (50) 2 x 8 =   (51) 8 x 8 =

(52) 2 x 7 =   (53) 8 x 12 =   (54) 9 x 10 =

(55) 4 x 4 =   (56) 11 x 10 =   (57) 8 x 5 =

(58) 9 x 12 =   (59) 8 x 11 =   (60) 5 x 7 =

(1) 7 x 8 =

(2) 1 x 6 =

(3) 5 x 6 =

(4) 10 x 4 =

(5) 1 x 6 =

(6) 7 x 4 =

(7) 1 x 5 =

(8) 2 x 11 =

(9) 5 x 10 =

(10) 6 x 2 =

(11) 3 x 2 =

(12) 1 x 12 =

(13) 10 x 3 =

(14) 1 x 1 =

(15) 1 x 3 =

(16) 8 x 5 =

(17) 1 x 11 =

(18) 8 x 10 =

(19) 9 x 10 =

(20) 6 x 1 =

(21) 5 x 9 =

(22) 10 x 6 =

(23) 10 x 7 =

(24) 4 x 1 =

(25) 9 x 5 =

(26) 4 x 10 =

(27) 12 x 12 =

(28) 3 x 5 =

(29) 6 x 2 =

(30) 3 x 12 =

(31) 6 x 9 =

(32) 4 x 7 =

(33) 8 x 4 =

(34) 8 x 10 =

(35) 4 x 7 =

(36) 7 x 7 =

(37) 11 x 1 =

(38) 2 x 3 =

(39) 12 x 8 =

(40) 3 x 9 =

(41) 3 x 2 =

(42) 12 x 7 =

(43) 9 x 12 =

(44) 4 x 5 =

(45) 12 x 7 =

(46) 4 x 12 =

(47) 10 x 2 =

(48) 3 x 1 =

(49) 10 x 11 =

(50) 12 x 4 =

(51) 2 x 4 =

(52) 4 x 8 =

(53) 3 x 2 =

(54) 6 x 10 =

(55) 6 x 9 =

(56) 6 x 5 =

(57) 7 x 8 =

(58) 3 x 5 =

(59) 4 x 4 =

(60) 6 x 9 =

(1) $9 \times 5 =$

(2) $12 \times 4 =$

(3) $3 \times 4 =$

(4) $2 \times 6 =$

(5) $8 \times 3 =$

(6) $1 \times 5 =$

(7) $3 \times 6 =$

(8) $5 \times 10 =$

(9) $9 \times 7 =$

(10) $7 \times 11 =$

(11) $10 \times 10 =$

(12) $3 \times 5 =$

(13) $12 \times 4 =$

(14) $3 \times 5 =$

(15) $12 \times 3 =$

(16) $2 \times 7 =$

(17) $5 \times 6 =$

(18) $11 \times 7 =$

(19) $6 \times 3 =$

(20) $10 \times 2 =$

(21) $8 \times 11 =$

(22) $6 \times 4 =$

(23) $9 \times 5 =$

(24) $3 \times 3 =$

(25) $5 \times 11 =$

(26) $8 \times 9 =$

(27) $12 \times 3 =$

(28) $12 \times 5 =$

(29) $1 \times 2 =$

(30) $12 \times 11 =$

(31) $1 \times 12 =$

(32) $12 \times 6 =$

(33) $12 \times 3 =$

(34) $7 \times 3 =$

(35) $6 \times 5 =$

(36) $2 \times 6 =$

(37) $10 \times 12 =$

(38) $12 \times 7 =$

(39) $10 \times 11 =$

(40) $8 \times 9 =$

(41) $2 \times 5 =$

(42) $2 \times 6 =$

(43) $12 \times 6 =$

(44) $6 \times 1 =$

(45) $12 \times 4 =$

(46) $6 \times 4 =$

(47) $11 \times 1 =$

(48) $5 \times 6 =$

(49) $4 \times 1 =$

(50) $11 \times 11 =$

(51) $7 \times 11 =$

(52) $8 \times 1 =$

(53) $11 \times 3 =$

(54) $10 \times 6 =$

(55) $8 \times 5 =$

(56) $11 \times 7 =$

(57) $4 \times 7 =$

(58) $2 \times 3 =$

(59) $7 \times 11 =$

(60) $12 \times 1 =$

(1) 3 x 4 =

(2) 1 x 6 =

(3) 6 x 11 =

(4) 1 x 1 =

(5) 9 x 12 =

(6) 2 x 11 =

(7) 7 x 1 =

(8) 11 x 12 =

(9) 7 x 2 =

(10) 3 x 11 =

(11) 1 x 8 =

(12) 10 x 2 =

(13) 8 x 8 =

(14) 11 x 4 =

(15) 6 x 10 =

(16) 1 x 6 =

(17) 6 x 6 =

(18) 10 x 11 =

(19) 8 x 4 =

(20) 12 x 9 =

(21) 11 x 7 =

(22) 4 x 9 =

(23) 9 x 1 =

(24) 11 x 6 =

(25) 1 x 8 =

(26) 10 x 8 =

(27) 2 x 11 =

(28) 9 x 1 =

(29) 10 x 1 =

(30) 11 x 3 =

(31) 9 x 7 =

(32) 10 x 1 =

(33) 5 x 8 =

(34) 2 x 11 =

(35) 9 x 7 =

(36) 6 x 8 =

(37) 6 x 9 =

(38) 4 x 5 =

(39) 9 x 11 =

(40) 5 x 2 =

(41) 5 x 6 =

(42) 3 x 2 =

(43) 5 x 9 =

(44) 8 x 4 =

(45) 11 x 5 =

(46) 10 x 5 =

(47) 5 x 1 =

(48) 5 x 6 =

(49) 3 x 1 =

(50) 12 x 11 =

(51) 3 x 3 =

(52) 8 x 2 =

(53) 5 x 3 =

(54) 1 x 12 =

(55) 1 x 6 =

(56) 9 x 2 =

(57) 5 x 9 =

(58) 11 x 9 =

(59) 5 x 3 =

(60) 4 x 9 =

(1) 6 x 12 =

(2) 12 x 5 =

(3) 7 x 7 =

(4) 1 x 12 =

(5) 10 x 12 =

(6) 7 x 3 =

(7) 5 x 12 =

(8) 9 x 6 =

(9) 10 x 6 =

(10) 10 x 3 =

(11) 3 x 7 =

(12) 10 x 3 =

(13) 1 x 10 =

(14) 11 x 5 =

(15) 2 x 12 =

(16) 1 x 11 =

(17) 2 x 8 =

(18) 4 x 12 =

(19) 10 x 12 =

(20) 11 x 12 =

(21) 3 x 3 =

(22) 6 x 1 =

(23) 7 x 5 =

(24) 5 x 9 =

(25) 1 x 2 =

(26) 1 x 8 =

(27) 12 x 4 =

(28) 9 x 11 =

(29) 2 x 4 =

(30) 10 x 12 =

(31) 8 x 1 =

(32) 11 x 9 =

(33) 9 x 3 =

(34) 1 x 2 =

(35) 11 x 9 =

(36) 2 x 4 =

(37) 8 x 4 =

(38) 7 x 1 =

(39) 2 x 4 =

(40) 2 x 4 =

(41) 10 x 6 =

(42) 1 x 4 =

(43) 7 x 9 =

(44) 6 x 3 =

(45) 2 x 3 =

(46) 1 x 4 =

(47) 11 x 7 =

(48) 6 x 3 =

(49) 8 x 1 =

(50) 9 x 1 =

(51) 4 x 12 =

(52) 5 x 12 =

(53) 2 x 2 =

(54) 10 x 11 =

(55) 8 x 5 =

(56) 1 x 12 =

(57) 4 x 8 =

(58) 12 x 4 =

(59) 6 x 3 =

(60) 2 x 8 =

(1) $9 \times 7 =$ 　　(2) $8 \times 11 =$ 　　(3) $5 \times 2 =$

(4) $12 \times 2 =$ 　　(5) $3 \times 7 =$ 　　(6) $10 \times 12 =$

(7) $9 \times 11 =$ 　　(8) $12 \times 12 =$ 　　(9) $7 \times 6 =$

(10) $6 \times 9 =$ 　　(11) $3 \times 5 =$ 　　(12) $6 \times 1 =$

(13) $6 \times 10 =$ 　　(14) $8 \times 7 =$ 　　(15) $7 \times 4 =$

(16) $6 \times 8 =$ 　　(17) $11 \times 5 =$ 　　(18) $4 \times 6 =$

(19) $6 \times 8 =$ 　　(20) $8 \times 11 =$ 　　(21) $12 \times 7 =$

(22) $7 \times 8 =$ 　　(23) $12 \times 10 =$ 　　(24) $12 \times 3 =$

(25) $11 \times 11 =$ 　　(26) $6 \times 11 =$ 　　(27) $2 \times 10 =$

(28) $10 \times 1 =$ 　　(29) $10 \times 9 =$ 　　(30) $4 \times 3 =$

(31) $6 \times 1 =$ 　　(32) $9 \times 12 =$ 　　(33) $12 \times 6 =$

(34) $12 \times 7 =$ 　　(35) $12 \times 11 =$ 　　(36) $8 \times 11 =$

(37) $12 \times 12 =$ 　　(38) $2 \times 11 =$ 　　(39) $6 \times 3 =$

(40) $6 \times 6 =$ 　　(41) $5 \times 8 =$ 　　(42) $2 \times 9 =$

(43) $12 \times 2 =$ 　　(44) $5 \times 9 =$ 　　(45) $7 \times 1 =$

(46) $12 \times 6 =$ 　　(47) $7 \times 11 =$ 　　(48) $6 \times 9 =$

(49) $6 \times 1 =$ 　　(50) $8 \times 9 =$ 　　(51) $4 \times 4 =$

(52) $12 \times 11 =$ 　　(53) $12 \times 7 =$ 　　(54) $7 \times 1 =$

(55) $10 \times 1 =$ 　　(56) $4 \times 10 =$ 　　(57) $4 \times 8 =$

(58) $6 \times 3 =$ 　　(59) $5 \times 1 =$ 　　(60) $8 \times 2 =$

(1) 2 x 8 =

(2) 2 x 9 =

(3) 10 x 4 =

(4) 3 x 6 =

(5) 12 x 4 =

(6) 3 x 7 =

(7) 8 x 3 =

(8) 7 x 6 =

(9) 2 x 7 =

(10) 9 x 1 =

(11) 11 x 5 =

(12) 2 x 3 =

(13) 3 x 8 =

(14) 4 x 1 =

(15) 8 x 11 =

(16) 2 x 6 =

(17) 4 x 9 =

(18) 1 x 5 =

(19) 6 x 4 =

(20) 2 x 4 =

(21) 6 x 4 =

(22) 11 x 5 =

(23) 1 x 10 =

(24) 2 x 7 =

(25) 8 x 6 =

(26) 3 x 3 =

(27) 4 x 8 =

(28) 12 x 5 =

(29) 8 x 2 =

(30) 4 x 7 =

(31) 4 x 9 =

(32) 9 x 2 =

(33) 4 x 11 =

(34) 2 x 7 =

(35) 6 x 4 =

(36) 6 x 1 =

(37) 1 x 5 =

(38) 5 x 2 =

(39) 9 x 6 =

(40) 2 x 6 =

(41) 6 x 2 =

(42) 5 x 7 =

(43) 10 x 2 =

(44) 4 x 6 =

(45) 5 x 11 =

(46) 4 x 12 =

(47) 6 x 10 =

(48) 1 x 4 =

(49) 6 x 4 =

(50) 5 x 5 =

(51) 8 x 5 =

(52) 7 x 4 =

(53) 3 x 7 =

(54) 5 x 6 =

(55) 6 x 12 =

(56) 2 x 8 =

(57) 10 x 2 =

(58) 8 x 2 =

(59) 3 x 5 =

(60) 8 x 2 =

Day:	14	Date:		Score:	/60
Name:		Time:	:	Rating:	☆☆☆☆☆☆

(1) 2 x 7 =

(2) 11 x 2 =

(3) 11 x 8 =

(4) 2 x 12 =

(5) 8 x 12 =

(6) 1 x 6 =

(7) 7 x 7 =

(8) 10 x 8 =

(9) 4 x 7 =

(10) 2 x 5 =

(11) 7 x 12 =

(12) 12 x 2 =

(13) 12 x 6 =

(14) 3 x 5 =

(15) 9 x 10 =

(16) 5 x 9 =

(17) 12 x 7 =

(18) 9 x 12 =

(19) 12 x 5 =

(20) 2 x 4 =

(21) 5 x 11 =

(22) 4 x 8 =

(23) 4 x 3 =

(24) 8 x 4 =

(25) 2 x 10 =

(26) 3 x 11 =

(27) 12 x 9 =

(28) 10 x 11 =

(29) 9 x 4 =

(30) 10 x 11 =

(31) 6 x 2 =

(32) 2 x 2 =

(33) 8 x 12 =

(34) 8 x 9 =

(35) 3 x 3 =

(36) 12 x 3 =

(37) 6 x 7 =

(38) 3 x 6 =

(39) 4 x 4 =

(40) 2 x 11 =

(41) 8 x 11 =

(42) 4 x 3 =

(43) 7 x 10 =

(44) 10 x 11 =

(45) 10 x 6 =

(46) 1 x 10 =

(47) 7 x 4 =

(48) 11 x 1 =

(49) 7 x 9 =

(50) 10 x 8 =

(51) 4 x 3 =

(52) 8 x 8 =

(53) 7 x 10 =

(54) 4 x 10 =

(55) 9 x 7 =

(56) 6 x 8 =

(57) 12 x 8 =

(58) 1 x 2 =

(59) 3 x 11 =

(60) 9 x 1 =

(1) 8 x 10 =

(2) 10 x 3 =

(3) 1 x 7 =

(4) 7 x 8 =

(5) 12 x 6 =

(6) 4 x 7 =

(7) 12 x 7 =

(8) 9 x 4 =

(9) 11 x 3 =

(10) 1 x 12 =

(11) 10 x 7 =

(12) 11 x 9 =

(13) 7 x 3 =

(14) 5 x 3 =

(15) 11 x 11 =

(16) 1 x 10 =

(17) 9 x 11 =

(18) 7 x 12 =

(19) 5 x 9 =

(20) 6 x 10 =

(21) 9 x 2 =

(22) 6 x 3 =

(23) 11 x 6 =

(24) 2 x 5 =

(25) 11 x 11 =

(26) 10 x 10 =

(27) 3 x 11 =

(28) 9 x 12 =

(29) 5 x 10 =

(30) 3 x 2 =

(31) 4 x 1 =

(32) 3 x 10 =

(33) 12 x 6 =

(34) 9 x 1 =

(35) 6 x 5 =

(36) 4 x 9 =

(37) 11 x 9 =

(38) 8 x 5 =

(39) 11 x 7 =

(40) 11 x 1 =

(41) 11 x 6 =

(42) 8 x 2 =

(43) 10 x 9 =

(44) 6 x 10 =

(45) 12 x 3 =

(46) 2 x 9 =

(47) 8 x 11 =

(48) 4 x 10 =

(49) 4 x 2 =

(50) 8 x 7 =

(51) 8 x 11 =

(52) 7 x 4 =

(53) 4 x 4 =

(54) 12 x 6 =

(55) 12 x 9 =

(56) 1 x 9 =

(57) 7 x 3 =

(58) 5 x 11 =

(59) 3 x 9 =

(60) 6 x 8 =

(1) 5 x 9 =
(2) 7 x 10 =
(3) 7 x 11 =

(4) 2 x 9 =
(5) 1 x 4 =
(6) 1 x 6 =

(7) 9 x 8 =
(8) 12 x 10 =
(9) 9 x 8 =

(10) 9 x 8 =
(11) 1 x 5 =
(12) 12 x 5 =

(13) 3 x 7 =
(14) 4 x 6 =
(15) 5 x 6 =

(16) 7 x 11 =
(17) 5 x 5 =
(18) 11 x 8 =

(19) 8 x 8 =
(20) 1 x 9 =
(21) 7 x 8 =

(22) 2 x 4 =
(23) 7 x 4 =
(24) 8 x 1 =

(25) 7 x 3 =
(26) 9 x 10 =
(27) 4 x 5 =

(28) 1 x 12 =
(29) 11 x 6 =
(30) 9 x 5 =

(31) 4 x 1 =
(32) 8 x 1 =
(33) 11 x 5 =

(34) 4 x 7 =
(35) 5 x 5 =
(36) 1 x 12 =

(37) 7 x 6 =
(38) 7 x 2 =
(39) 6 x 7 =

(40) 8 x 4 =
(41) 8 x 5 =
(42) 12 x 5 =

(43) 10 x 12 =
(44) 11 x 4 =
(45) 11 x 7 =

(46) 10 x 5 =
(47) 12 x 8 =
(48) 5 x 12 =

(49) 5 x 12 =
(50) 2 x 7 =
(51) 9 x 6 =

(52) 7 x 12 =
(53) 7 x 3 =
(54) 3 x 7 =

(55) 4 x 9 =
(56) 8 x 12 =
(57) 9 x 2 =

(58) 6 x 8 =
(59) 10 x 8 =
(60) 12 x 8 =

(1) 10 x 2 =

(2) 10 x 1 =

(3) 9 x 4 =

(4) 2 x 2 =

(5) 2 x 6 =

(6) 12 x 7 =

(7) 2 x 5 =

(8) 1 x 5 =

(9) 6 x 2 =

(10) 9 x 4 =

(11) 4 x 3 =

(12) 8 x 3 =

(13) 3 x 8 =

(14) 3 x 12 =

(15) 9 x 2 =

(16) 8 x 5 =

(17) 4 x 9 =

(18) 5 x 6 =

(19) 1 x 12 =

(20) 7 x 7 =

(21) 11 x 3 =

(22) 12 x 1 =

(23) 1 x 11 =

(24) 7 x 12 =

(25) 11 x 2 =

(26) 2 x 10 =

(27) 11 x 10 =

(28) 9 x 3 =

(29) 10 x 6 =

(30) 1 x 4 =

(31) 11 x 4 =

(32) 3 x 10 =

(33) 3 x 1 =

(34) 5 x 7 =

(35) 9 x 3 =

(36) 2 x 8 =

(37) 6 x 11 =

(38) 6 x 4 =

(39) 5 x 8 =

(40) 6 x 11 =

(41) 6 x 9 =

(42) 4 x 1 =

(43) 10 x 8 =

(44) 4 x 11 =

(45) 9 x 7 =

(46) 8 x 4 =

(47) 3 x 4 =

(48) 2 x 11 =

(49) 11 x 2 =

(50) 6 x 6 =

(51) 10 x 2 =

(52) 11 x 11 =

(53) 7 x 1 =

(54) 5 x 5 =

(55) 11 x 1 =

(56) 8 x 2 =

(57) 5 x 3 =

(58) 2 x 2 =

(59) 10 x 12 =

(60) 11 x 8 =

(1) $12 \times 7 =$

(2) $9 \times 5 =$

(3) $9 \times 3 =$

(4) $12 \times 9 =$

(5) $10 \times 5 =$

(6) $6 \times 6 =$

(7) $9 \times 7 =$

(8) $10 \times 4 =$

(9) $6 \times 7 =$

(10) $5 \times 10 =$

(11) $12 \times 7 =$

(12) $3 \times 7 =$

(13) $5 \times 9 =$

(14) $10 \times 1 =$

(15) $12 \times 9 =$

(16) $6 \times 2 =$

(17) $10 \times 7 =$

(18) $10 \times 2 =$

(19) $4 \times 9 =$

(20) $10 \times 6 =$

(21) $3 \times 4 =$

(22) $6 \times 1 =$

(23) $1 \times 3 =$

(24) $4 \times 3 =$

(25) $4 \times 2 =$

(26) $9 \times 1 =$

(27) $9 \times 10 =$

(28) $7 \times 5 =$

(29) $3 \times 3 =$

(30) $9 \times 7 =$

(31) $10 \times 7 =$

(32) $11 \times 4 =$

(33) $6 \times 11 =$

(34) $12 \times 10 =$

(35) $6 \times 11 =$

(36) $6 \times 3 =$

(37) $9 \times 2 =$

(38) $4 \times 6 =$

(39) $7 \times 6 =$

(40) $2 \times 7 =$

(41) $8 \times 10 =$

(42) $3 \times 10 =$

(43) $10 \times 1 =$

(44) $4 \times 10 =$

(45) $6 \times 6 =$

(46) $10 \times 12 =$

(47) $5 \times 2 =$

(48) $9 \times 11 =$

(49) $1 \times 6 =$

(50) $7 \times 10 =$

(51) $9 \times 9 =$

(52) $3 \times 12 =$

(53) $5 \times 10 =$

(54) $10 \times 9 =$

(55) $4 \times 12 =$

(56) $6 \times 6 =$

(57) $1 \times 3 =$

(58) $5 \times 11 =$

(59) $4 \times 6 =$

(60) $7 \times 12 =$

(1) $12 \times 6 =$ (2) $6 \times 4 =$ (3) $1 \times 11 =$

(4) $12 \times 7 =$ (5) $1 \times 1 =$ (6) $5 \times 1 =$

(7) $7 \times 9 =$ (8) $4 \times 8 =$ (9) $11 \times 2 =$

(10) $12 \times 4 =$ (11) $5 \times 12 =$ (12) $10 \times 9 =$

(13) $11 \times 11 =$ (14) $3 \times 10 =$ (15) $5 \times 4 =$

(16) $2 \times 12 =$ (17) $4 \times 6 =$ (18) $3 \times 10 =$

(19) $8 \times 6 =$ (20) $1 \times 1 =$ (21) $6 \times 7 =$

(22) $7 \times 12 =$ (23) $4 \times 10 =$ (24) $8 \times 9 =$

(25) $9 \times 7 =$ (26) $8 \times 6 =$ (27) $4 \times 5 =$

(28) $9 \times 3 =$ (29) $5 \times 8 =$ (30) $10 \times 9 =$

(31) $10 \times 8 =$ (32) $1 \times 4 =$ (33) $6 \times 12 =$

(34) $4 \times 5 =$ (35) $8 \times 4 =$ (36) $10 \times 5 =$

(37) $7 \times 7 =$ (38) $11 \times 7 =$ (39) $10 \times 5 =$

(40) $6 \times 3 =$ (41) $3 \times 8 =$ (42) $11 \times 6 =$

(43) $9 \times 9 =$ (44) $7 \times 4 =$ (45) $3 \times 4 =$

(46) $11 \times 11 =$ (47) $3 \times 6 =$ (48) $2 \times 2 =$

(49) $10 \times 4 =$ (50) $6 \times 7 =$ (51) $3 \times 1 =$

(52) $6 \times 4 =$ (53) $9 \times 10 =$ (54) $12 \times 8 =$

(55) $12 \times 3 =$ (56) $12 \times 6 =$ (57) $3 \times 8 =$

(58) $9 \times 11 =$ (59) $11 \times 11 =$ (60) $7 \times 7 =$

(1) 9 x 7 =

(2) 4 x 11 =

(3) 3 x 7 =

(4) 2 x 11 =

(5) 6 x 4 =

(6) 3 x 4 =

(7) 12 x 8 =

(8) 3 x 9 =

(9) 8 x 8 =

(10) 7 x 11 =

(11) 2 x 3 =

(12) 11 x 6 =

(13) 9 x 1 =

(14) 1 x 9 =

(15) 4 x 6 =

(16) 1 x 10 =

(17) 4 x 4 =

(18) 10 x 3 =

(19) 2 x 12 =

(20) 6 x 4 =

(21) 11 x 9 =

(22) 3 x 7 =

(23) 5 x 1 =

(24) 2 x 10 =

(25) 10 x 6 =

(26) 8 x 2 =

(27) 12 x 9 =

(28) 1 x 9 =

(29) 2 x 7 =

(30) 2 x 11 =

(31) 9 x 8 =

(32) 12 x 1 =

(33) 6 x 10 =

(34) 4 x 7 =

(35) 12 x 8 =

(36) 1 x 7 =

(37) 6 x 3 =

(38) 7 x 9 =

(39) 8 x 10 =

(40) 11 x 9 =

(41) 9 x 9 =

(42) 12 x 1 =

(43) 5 x 11 =

(44) 2 x 5 =

(45) 11 x 3 =

(46) 5 x 12 =

(47) 5 x 4 =

(48) 4 x 4 =

(49) 4 x 11 =

(50) 12 x 1 =

(51) 4 x 6 =

(52) 2 x 5 =

(53) 5 x 1 =

(54) 1 x 6 =

(55) 11 x 6 =

(56) 2 x 1 =

(57) 3 x 3 =

(58) 4 x 7 =

(59) 2 x 9 =

(60) 7 x 11 =

(1) $5 \times 12 =$

(2) $8 \times 12 =$

(3) $11 \times 4 =$

(4) $3 \times 12 =$

(5) $8 \times 4 =$

(6) $2 \times 5 =$

(7) $3 \times 4 =$

(8) $6 \times 8 =$

(9) $10 \times 6 =$

(10) $8 \times 12 =$

(11) $10 \times 8 =$

(12) $5 \times 7 =$

(13) $7 \times 10 =$

(14) $8 \times 3 =$

(15) $1 \times 12 =$

(16) $9 \times 2 =$

(17) $8 \times 6 =$

(18) $2 \times 7 =$

(19) $11 \times 3 =$

(20) $5 \times 9 =$

(21) $11 \times 5 =$

(22) $9 \times 1 =$

(23) $7 \times 1 =$

(24) $11 \times 12 =$

(25) $2 \times 9 =$

(26) $12 \times 7 =$

(27) $11 \times 2 =$

(28) $2 \times 10 =$

(29) $5 \times 9 =$

(30) $2 \times 2 =$

(31) $9 \times 7 =$

(32) $8 \times 7 =$

(33) $6 \times 2 =$

(34) $11 \times 8 =$

(35) $10 \times 8 =$

(36) $1 \times 1 =$

(37) $11 \times 8 =$

(38) $9 \times 9 =$

(39) $11 \times 2 =$

(40) $8 \times 4 =$

(41) $9 \times 7 =$

(42) $3 \times 1 =$

(43) $8 \times 11 =$

(44) $1 \times 10 =$

(45) $8 \times 3 =$

(46) $2 \times 8 =$

(47) $7 \times 4 =$

(48) $5 \times 5 =$

(49) $1 \times 12 =$

(50) $11 \times 2 =$

(51) $2 \times 5 =$

(52) $2 \times 4 =$

(53) $1 \times 12 =$

(54) $2 \times 12 =$

(55) $6 \times 3 =$

(56) $10 \times 10 =$

(57) $3 \times 5 =$

(58) $4 \times 12 =$

(59) $11 \times 2 =$

(60) $10 \times 4 =$

(1) $12 \times 10 =$ (2) $3 \times 6 =$ (3) $3 \times 5 =$

(4) $4 \times 10 =$ (5) $9 \times 2 =$ (6) $3 \times 9 =$

(7) $2 \times 1 =$ (8) $7 \times 5 =$ (9) $7 \times 3 =$

(10) $7 \times 6 =$ (11) $6 \times 5 =$ (12) $1 \times 12 =$

(13) $9 \times 8 =$ (14) $6 \times 5 =$ (15) $1 \times 4 =$

(16) $4 \times 11 =$ (17) $5 \times 9 =$ (18) $6 \times 2 =$

(19) $6 \times 5 =$ (20) $6 \times 11 =$ (21) $3 \times 10 =$

(22) $8 \times 8 =$ (23) $7 \times 2 =$ (24) $7 \times 4 =$

(25) $12 \times 6 =$ (26) $7 \times 5 =$ (27) $4 \times 1 =$

(28) $6 \times 6 =$ (29) $9 \times 4 =$ (30) $6 \times 7 =$

(31) $4 \times 2 =$ (32) $9 \times 3 =$ (33) $7 \times 10 =$

(34) $11 \times 10 =$ (35) $5 \times 12 =$ (36) $1 \times 5 =$

(37) $5 \times 4 =$ (38) $7 \times 11 =$ (39) $10 \times 2 =$

(40) $10 \times 11 =$ (41) $4 \times 2 =$ (42) $12 \times 7 =$

(43) $9 \times 9 =$ (44) $11 \times 9 =$ (45) $4 \times 6 =$

(46) $4 \times 10 =$ (47) $8 \times 12 =$ (48) $12 \times 7 =$

(49) $7 \times 1 =$ (50) $3 \times 11 =$ (51) $8 \times 2 =$

(52) $8 \times 9 =$ (53) $9 \times 5 =$ (54) $1 \times 10 =$

(55) $6 \times 12 =$ (56) $10 \times 5 =$ (57) $2 \times 2 =$

(58) $11 \times 6 =$ (59) $6 \times 2 =$ (60) $5 \times 4 =$

(1) 8 x 8 =

(2) 2 x 12 =

(3) 5 x 1 =

(4) 8 x 9 =

(5) 1 x 8 =

(6) 12 x 3 =

(7) 1 x 1 =

(8) 7 x 1 =

(9) 3 x 8 =

(10) 10 x 6 =

(11) 2 x 4 =

(12) 5 x 5 =

(13) 10 x 7 =

(14) 4 x 12 =

(15) 4 x 7 =

(16) 3 x 4 =

(17) 4 x 6 =

(18) 11 x 4 =

(19) 8 x 7 =

(20) 10 x 3 =

(21) 5 x 9 =

(22) 8 x 12 =

(23) 7 x 7 =

(24) 10 x 4 =

(25) 9 x 7 =

(26) 2 x 8 =

(27) 3 x 1 =

(28) 7 x 11 =

(29) 10 x 7 =

(30) 5 x 7 =

(31) 1 x 1 =

(32) 3 x 11 =

(33) 12 x 3 =

(34) 1 x 6 =

(35) 1 x 3 =

(36) 12 x 5 =

(37) 5 x 12 =

(38) 12 x 6 =

(39) 4 x 12 =

(40) 4 x 10 =

(41) 8 x 10 =

(42) 5 x 7 =

(43) 3 x 7 =

(44) 9 x 5 =

(45) 10 x 8 =

(46) 2 x 9 =

(47) 6 x 6 =

(48) 11 x 4 =

(49) 6 x 10 =

(50) 2 x 8 =

(51) 1 x 10 =

(52) 3 x 1 =

(53) 7 x 12 =

(54) 2 x 7 =

(55) 5 x 8 =

(56) 8 x 9 =

(57) 4 x 8 =

(58) 4 x 5 =

(59) 9 x 3 =

(60) 1 x 2 =

Day:	24	Date:		Score:	/60
Name:		Time:	:	Rating:	☆☆☆☆☆

(1) $10 \times 9 =$

(2) $2 \times 1 =$

(3) $1 \times 1 =$

(4) $5 \times 5 =$

(5) $5 \times 9 =$

(6) $11 \times 5 =$

(7) $4 \times 6 =$

(8) $9 \times 6 =$

(9) $8 \times 6 =$

(10) $12 \times 1 =$

(11) $7 \times 8 =$

(12) $5 \times 11 =$

(13) $10 \times 8 =$

(14) $7 \times 6 =$

(15) $1 \times 11 =$

(16) $2 \times 5 =$

(17) $4 \times 12 =$

(18) $10 \times 4 =$

(19) $5 \times 11 =$

(20) $4 \times 3 =$

(21) $4 \times 11 =$

(22) $7 \times 2 =$

(23) $4 \times 12 =$

(24) $8 \times 3 =$

(25) $9 \times 8 =$

(26) $4 \times 1 =$

(27) $12 \times 8 =$

(28) $5 \times 8 =$

(29) $4 \times 11 =$

(30) $2 \times 1 =$

(31) $12 \times 2 =$

(32) $1 \times 2 =$

(33) $5 \times 6 =$

(34) $1 \times 8 =$

(35) $9 \times 1 =$

(36) $3 \times 8 =$

(37) $6 \times 12 =$

(38) $3 \times 1 =$

(39) $1 \times 5 =$

(40) $12 \times 2 =$

(41) $6 \times 9 =$

(42) $5 \times 11 =$

(43) $2 \times 7 =$

(44) $12 \times 3 =$

(45) $7 \times 12 =$

(46) $11 \times 12 =$

(47) $1 \times 1 =$

(48) $7 \times 10 =$

(49) $9 \times 2 =$

(50) $6 \times 5 =$

(51) $12 \times 10 =$

(52) $6 \times 3 =$

(53) $5 \times 3 =$

(54) $10 \times 3 =$

(55) $1 \times 3 =$

(56) $3 \times 3 =$

(57) $4 \times 6 =$

(58) $9 \times 2 =$

(59) $1 \times 4 =$

(60) $7 \times 9 =$

(1) $5 \times 12 =$

(2) $3 \times 11 =$

(3) $2 \times 4 =$

(4) $5 \times 4 =$

(5) $3 \times 10 =$

(6) $10 \times 4 =$

(7) $2 \times 6 =$

(8) $7 \times 3 =$

(9) $4 \times 12 =$

(10) $7 \times 5 =$

(11) $9 \times 6 =$

(12) $1 \times 10 =$

(13) $10 \times 1 =$

(14) $10 \times 12 =$

(15) $3 \times 10 =$

(16) $1 \times 5 =$

(17) $1 \times 10 =$

(18) $5 \times 4 =$

(19) $5 \times 6 =$

(20) $9 \times 2 =$

(21) $2 \times 9 =$

(22) $6 \times 6 =$

(23) $1 \times 1 =$

(24) $1 \times 5 =$

(25) $1 \times 11 =$

(26) $4 \times 4 =$

(27) $9 \times 3 =$

(28) $5 \times 4 =$

(29) $5 \times 7 =$

(30) $1 \times 2 =$

(31) $12 \times 4 =$

(32) $9 \times 4 =$

(33) $1 \times 10 =$

(34) $4 \times 4 =$

(35) $8 \times 2 =$

(36) $9 \times 9 =$

(37) $4 \times 2 =$

(38) $6 \times 2 =$

(39) $11 \times 3 =$

(40) $7 \times 12 =$

(41) $9 \times 12 =$

(42) $9 \times 4 =$

(43) $6 \times 6 =$

(44) $10 \times 1 =$

(45) $6 \times 12 =$

(46) $9 \times 5 =$

(47) $1 \times 8 =$

(48) $3 \times 6 =$

(49) $9 \times 10 =$

(50) $4 \times 4 =$

(51) $10 \times 9 =$

(52) $6 \times 5 =$

(53) $4 \times 1 =$

(54) $9 \times 9 =$

(55) $1 \times 9 =$

(56) $6 \times 1 =$

(57) $3 \times 11 =$

(58) $1 \times 12 =$

(59) $1 \times 5 =$

(60) $3 \times 2 =$

Day:	26	Date:		Score:	/60
Name:		Time:	:	Rating:	☆☆☆☆☆☆

(1) 2 x 3 =   (2) 4 x 10 =   (3) 7 x 1 =

(4) 3 x 1 =   (5) 1 x 11 =   (6) 2 x 2 =

(7) 3 x 4 =   (8) 4 x 10 =   (9) 2 x 2 =

(10) 4 x 6 =   (11) 11 x 9 =   (12) 8 x 1 =

(13) 12 x 1 =   (14) 7 x 6 =   (15) 1 x 7 =

(16) 12 x 10 =   (17) 6 x 4 =   (18) 5 x 2 =

(19) 12 x 11 =   (20) 9 x 2 =   (21) 2 x 11 =

(22) 12 x 8 =   (23) 10 x 9 =   (24) 3 x 1 =

(25) 3 x 2 =   (26) 6 x 8 =   (27) 3 x 11 =

(28) 10 x 5 =   (29) 10 x 12 =   (30) 11 x 6 =

(31) 5 x 7 =   (32) 5 x 10 =   (33) 6 x 9 =

(34) 3 x 4 =   (35) 10 x 4 =   (36) 8 x 7 =

(37) 8 x 10 =   (38) 6 x 8 =   (39) 12 x 9 =

(40) 5 x 10 =   (41) 4 x 9 =   (42) 10 x 1 =

(43) 3 x 9 =   (44) 5 x 3 =   (45) 12 x 10 =

(46) 10 x 3 =   (47) 7 x 6 =   (48) 6 x 6 =

(49) 7 x 9 =   (50) 12 x 1 =   (51) 1 x 9 =

(52) 4 x 2 =   (53) 2 x 2 =   (54) 11 x 5 =

(55) 9 x 4 =   (56) 5 x 1 =   (57) 2 x 6 =

(58) 2 x 8 =   (59) 2 x 5 =   (60) 4 x 10 =

(1) 9 x 6 =

(2) 7 x 7 =

(3) 6 x 10 =

(4) 4 x 12 =

(5) 8 x 6 =

(6) 5 x 2 =

(7) 3 x 12 =

(8) 9 x 1 =

(9) 7 x 3 =

(10) 3 x 7 =

(11) 1 x 5 =

(12) 11 x 6 =

(13) 3 x 9 =

(14) 3 x 8 =

(15) 12 x 1 =

(16) 12 x 10 =

(17) 2 x 7 =

(18) 12 x 1 =

(19) 1 x 3 =

(20) 11 x 11 =

(21) 7 x 12 =

(22) 9 x 8 =

(23) 6 x 12 =

(24) 10 x 1 =

(25) 10 x 5 =

(26) 3 x 4 =

(27) 9 x 11 =

(28) 6 x 8 =

(29) 9 x 6 =

(30) 1 x 3 =

(31) 3 x 11 =

(32) 12 x 11 =

(33) 3 x 3 =

(34) 8 x 1 =

(35) 5 x 8 =

(36) 10 x 4 =

(37) 12 x 6 =

(38) 2 x 7 =

(39) 9 x 1 =

(40) 10 x 9 =

(41) 10 x 1 =

(42) 9 x 2 =

(43) 11 x 1 =

(44) 7 x 5 =

(45) 2 x 7 =

(46) 11 x 5 =

(47) 4 x 6 =

(48) 1 x 6 =

(49) 5 x 12 =

(50) 3 x 5 =

(51) 4 x 4 =

(52) 4 x 12 =

(53) 6 x 9 =

(54) 6 x 8 =

(55) 8 x 6 =

(56) 2 x 11 =

(57) 9 x 2 =

(58) 9 x 7 =

(59) 12 x 7 =

(60) 9 x 3 =

(1) $2 \times 12 =$

(2) $1 \times 11 =$

(3) $10 \times 1 =$

(4) $9 \times 3 =$

(5) $11 \times 11 =$

(6) $3 \times 4 =$

(7) $2 \times 6 =$

(8) $6 \times 6 =$

(9) $7 \times 4 =$

(10) $1 \times 5 =$

(11) $12 \times 4 =$

(12) $12 \times 5 =$

(13) $3 \times 12 =$

(14) $10 \times 1 =$

(15) $6 \times 11 =$

(16) $3 \times 4 =$

(17) $7 \times 9 =$

(18) $2 \times 10 =$

(19) $12 \times 4 =$

(20) $11 \times 1 =$

(21) $3 \times 6 =$

(22) $7 \times 9 =$

(23) $5 \times 2 =$

(24) $4 \times 2 =$

(25) $12 \times 9 =$

(26) $7 \times 2 =$

(27) $11 \times 2 =$

(28) $7 \times 8 =$

(29) $8 \times 2 =$

(30) $3 \times 2 =$

(31) $5 \times 4 =$

(32) $3 \times 11 =$

(33) $12 \times 8 =$

(34) $11 \times 7 =$

(35) $6 \times 10 =$

(36) $2 \times 7 =$

(37) $6 \times 4 =$

(38) $2 \times 6 =$

(39) $2 \times 9 =$

(40) $10 \times 4 =$

(41) $5 \times 3 =$

(42) $11 \times 6 =$

(43) $4 \times 7 =$

(44) $1 \times 1 =$

(45) $1 \times 12 =$

(46) $10 \times 6 =$

(47) $7 \times 2 =$

(48) $5 \times 6 =$

(49) $12 \times 11 =$

(50) $7 \times 5 =$

(51) $8 \times 3 =$

(52) $5 \times 1 =$

(53) $7 \times 6 =$

(54) $9 \times 8 =$

(55) $10 \times 11 =$

(56) $3 \times 6 =$

(57) $9 \times 5 =$

(58) $9 \times 3 =$

(59) $9 \times 6 =$

(60) $6 \times 7 =$

(1) 10 x 9 =

(2) 11 x 9 =

(3) 4 x 2 =

(4) 6 x 5 =

(5) 2 x 9 =

(6) 5 x 5 =

(7) 2 x 1 =

(8) 11 x 4 =

(9) 7 x 10 =

(10) 3 x 12 =

(11) 9 x 7 =

(12) 12 x 5 =

(13) 11 x 10 =

(14) 5 x 3 =

(15) 9 x 1 =

(16) 6 x 10 =

(17) 11 x 3 =

(18) 2 x 5 =

(19) 2 x 1 =

(20) 1 x 4 =

(21) 4 x 8 =

(22) 10 x 11 =

(23) 8 x 2 =

(24) 7 x 8 =

(25) 9 x 5 =

(26) 11 x 3 =

(27) 1 x 4 =

(28) 3 x 11 =

(29) 9 x 11 =

(30) 9 x 12 =

(31) 9 x 8 =

(32) 8 x 5 =

(33) 12 x 3 =

(34) 12 x 9 =

(35) 8 x 11 =

(36) 7 x 4 =

(37) 10 x 12 =

(38) 8 x 10 =

(39) 3 x 10 =

(40) 7 x 9 =

(41) 9 x 1 =

(42) 5 x 1 =

(43) 8 x 9 =

(44) 7 x 7 =

(45) 5 x 4 =

(46) 12 x 8 =

(47) 6 x 10 =

(48) 11 x 8 =

(49) 12 x 12 =

(50) 4 x 9 =

(51) 9 x 9 =

(52) 2 x 4 =

(53) 2 x 8 =

(54) 6 x 4 =

(55) 8 x 5 =

(56) 12 x 5 =

(57) 9 x 6 =

(58) 1 x 6 =

(59) 3 x 8 =

(60) 12 x 5 =

Day:	30	Date:		Score:	/60
Name:		Time:	:	Rating:	☆☆☆☆☆

(1) $8 \times 11 =$

(2) $3 \times 6 =$

(3) $3 \times 11 =$

(4) $10 \times 10 =$

(5) $2 \times 6 =$

(6) $4 \times 4 =$

(7) $4 \times 11 =$

(8) $5 \times 10 =$

(9) $3 \times 4 =$

(10) $7 \times 1 =$

(11) $6 \times 7 =$

(12) $1 \times 7 =$

(13) $8 \times 3 =$

(14) $1 \times 1 =$

(15) $9 \times 10 =$

(16) $2 \times 6 =$

(17) $9 \times 6 =$

(18) $5 \times 7 =$

(19) $4 \times 8 =$

(20) $4 \times 3 =$

(21) $1 \times 8 =$

(22) $6 \times 5 =$

(23) $10 \times 6 =$

(24) $1 \times 4 =$

(25) $7 \times 7 =$

(26) $3 \times 3 =$

(27) $5 \times 1 =$

(28) $9 \times 11 =$

(29) $8 \times 3 =$

(30) $2 \times 1 =$

(31) $10 \times 9 =$

(32) $11 \times 12 =$

(33) $12 \times 11 =$

(34) $3 \times 2 =$

(35) $7 \times 6 =$

(36) $11 \times 12 =$

(37) $2 \times 8 =$

(38) $9 \times 4 =$

(39) $11 \times 10 =$

(40) $8 \times 6 =$

(41) $11 \times 1 =$

(42) $4 \times 1 =$

(43) $8 \times 5 =$

(44) $3 \times 3 =$

(45) $10 \times 10 =$

(46) $5 \times 2 =$

(47) $9 \times 10 =$

(48) $8 \times 12 =$

(49) $11 \times 8 =$

(50) $7 \times 11 =$

(51) $11 \times 3 =$

(52) $1 \times 5 =$

(53) $5 \times 3 =$

(54) $10 \times 8 =$

(55) $3 \times 5 =$

(56) $6 \times 12 =$

(57) $5 \times 5 =$

(58) $11 \times 1 =$

(59) $10 \times 5 =$

(60) $8 \times 1 =$

(1) 12 x 1 =   (2) 11 x 11 =   (3) 12 x 4 =

(4) 12 x 12 =   (5) 10 x 8 =   (6) 2 x 11 =

(7) 5 x 1 =   (8) 5 x 11 =   (9) 6 x 1 =

(10) 9 x 12 =   (11) 8 x 12 =   (12) 2 x 11 =

(13) 10 x 9 =   (14) 11 x 7 =   (15) 7 x 1 =

(16) 10 x 8 =   (17) 3 x 8 =   (18) 11 x 3 =

(19) 12 x 12 = (20) 11 x 12 =   (21) 4 x 5 =

(22) 11 x 5 =   (23) 6 x 2 =   (24) 11 x 5 =

(25) 8 x 10 =   (26) 7 x 10 =   (27) 10 x 8 =

(28) 11 x 9 =   (29) 4 x 2 =   (30) 3 x 11 =

(31) 5 x 7 =   (32) 3 x 7 =   (33) 2 x 5 =

(34) 9 x 12 =   (35) 10 x 11 =   (36) 12 x 12 =

(37) 12 x 11 = (38) 12 x 12 =   (39) 2 x 9 =

(40) 6 x 9 =   (41) 8 x 6 =   (42) 2 x 3 =

(43) 6 x 9 =   (44) 1 x 11 =   (45) 8 x 1 =

(46) 7 x 7 =   (47) 6 x 9 =   (48) 1 x 3 =

(49) 1 x 6 =   (50) 3 x 5 =   (51) 10 x 11 =

(52) 10 x 4 =   (53) 5 x 9 =   (54) 7 x 9 =

(55) 8 x 9 =   (56) 11 x 3 =   (57) 9 x 11 =

(58) 1 x 11 =   (59) 2 x 7 =   (60) 3 x 10 =

| Day: | 32 | Date: | | Score: | /60 |
| Name: | | Time: | : | Rating: | ☆☆☆☆☆ |

(1) 4 x 5 =

(2) 2 x 9 =

(3) 11 x 11 =

(4) 3 x 3 =

(5) 8 x 4 =

(6) 2 x 2 =

(7) 8 x 6 =

(8) 10 x 1 =

(9) 5 x 10 =

(10) 7 x 2 =

(11) 7 x 8 =

(12) 6 x 2 =

(13) 11 x 6 =

(14) 9 x 11 =

(15) 1 x 4 =

(16) 10 x 10 =

(17) 10 x 4 =

(18) 1 x 4 =

(19) 2 x 10 =

(20) 4 x 1 =

(21) 8 x 12 =

(22) 3 x 1 =

(23) 5 x 7 =

(24) 5 x 12 =

(25) 11 x 8 =

(26) 8 x 3 =

(27) 11 x 9 =

(28) 8 x 3 =

(29) 5 x 5 =

(30) 6 x 3 =

(31) 12 x 5 =

(32) 4 x 7 =

(33) 12 x 5 =

(34) 3 x 9 =

(35) 12 x 8 =

(36) 6 x 5 =

(37) 7 x 11 =

(38) 5 x 6 =

(39) 3 x 7 =

(40) 8 x 7 =

(41) 6 x 10 =

(42) 12 x 10 =

(43) 11 x 5 =

(44) 9 x 5 =

(45) 5 x 4 =

(46) 5 x 5 =

(47) 5 x 1 =

(48) 5 x 6 =

(49) 3 x 12 =

(50) 1 x 3 =

(51) 6 x 12 =

(52) 5 x 3 =

(53) 6 x 8 =

(54) 4 x 8 =

(55) 3 x 6 =

(56) 2 x 3 =

(57) 5 x 2 =

(58) 3 x 7 =

(59) 12 x 7 =

(60) 6 x 7 =

(1) $2 \times 4 =$

(2) $9 \times 11 =$

(3) $7 \times 1 =$

(4) $8 \times 10 =$

(5) $11 \times 1 =$

(6) $5 \times 5 =$

(7) $1 \times 8 =$

(8) $3 \times 5 =$

(9) $2 \times 12 =$

(10) $4 \times 3 =$

(11) $1 \times 2 =$

(12) $7 \times 7 =$

(13) $7 \times 2 =$

(14) $9 \times 6 =$

(15) $5 \times 2 =$

(16) $8 \times 9 =$

(17) $7 \times 12 =$

(18) $11 \times 10 =$

(19) $12 \times 9 =$

(20) $4 \times 7 =$

(21) $10 \times 12 =$

(22) $4 \times 3 =$

(23) $12 \times 9 =$

(24) $7 \times 5 =$

(25) $5 \times 7 =$

(26) $6 \times 8 =$

(27) $10 \times 6 =$

(28) $8 \times 8 =$

(29) $12 \times 8 =$

(30) $4 \times 8 =$

(31) $11 \times 11 =$

(32) $12 \times 3 =$

(33) $12 \times 9 =$

(34) $11 \times 9 =$

(35) $3 \times 4 =$

(36) $5 \times 8 =$

(37) $2 \times 9 =$

(38) $2 \times 8 =$

(39) $8 \times 2 =$

(40) $5 \times 11 =$

(41) $7 \times 7 =$

(42) $6 \times 11 =$

(43) $11 \times 12 =$

(44) $10 \times 10 =$

(45) $3 \times 4 =$

(46) $12 \times 1 =$

(47) $4 \times 8 =$

(48) $2 \times 3 =$

(49) $11 \times 7 =$

(50) $6 \times 10 =$

(51) $4 \times 4 =$

(52) $3 \times 5 =$

(53) $3 \times 8 =$

(54) $9 \times 5 =$

(55) $5 \times 8 =$

(56) $10 \times 9 =$

(57) $3 \times 8 =$

(58) $8 \times 5 =$

(59) $7 \times 9 =$

(60) $9 \times 8 =$

Day:	34	Date:		Score:	/60
Name:		Time:	:	Rating:	☆☆☆☆☆☆

(1) 2 x 7 =

(2) 2 x 4 =

(3) 3 x 1 =

(4) 9 x 12 =

(5) 7 x 1 =

(6) 12 x 4 =

(7) 10 x 7 =

(8) 8 x 3 =

(9) 12 x 9 =

(10) 12 x 3 =

(11) 6 x 2 =

(12) 8 x 11 =

(13) 3 x 8 =

(14) 9 x 11 =

(15) 1 x 3 =

(16) 1 x 4 =

(17) 8 x 8 =

(18) 8 x 10 =

(19) 2 x 5 =

(20) 1 x 2 =

(21) 9 x 1 =

(22) 11 x 1 =

(23) 10 x 3 =

(24) 10 x 6 =

(25) 11 x 7 =

(26) 6 x 2 =

(27) 10 x 7 =

(28) 1 x 7 =

(29) 1 x 8 =

(30) 7 x 9 =

(31) 1 x 6 =

(32) 2 x 11 =

(33) 12 x 9 =

(34) 6 x 11 =

(35) 7 x 2 =

(36) 8 x 6 =

(37) 2 x 1 =

(38) 7 x 10 =

(39) 3 x 2 =

(40) 1 x 6 =

(41) 10 x 2 =

(42) 2 x 3 =

(43) 6 x 1 =

(44) 3 x 9 =

(45) 4 x 1 =

(46) 8 x 4 =

(47) 2 x 11 =

(48) 4 x 3 =

(49) 1 x 3 =

(50) 1 x 2 =

(51) 5 x 11 =

(52) 4 x 3 =

(53) 1 x 9 =

(54) 1 x 8 =

(55) 4 x 6 =

(56) 8 x 10 =

(57) 4 x 11 =

(58) 11 x 4 =

(59) 1 x 1 =

(60) 4 x 2 =

(1) 1 x 1 =
(2) 3 x 5 =
(3) 2 x 11 =

(4) 12 x 12 =
(5) 5 x 10 =
(6) 9 x 7 =

(7) 2 x 5 =
(8) 8 x 5 =
(9) 1 x 2 =

(10) 5 x 8 =
(11) 6 x 10 =
(12) 4 x 7 =

(13) 9 x 7 =
(14) 6 x 4 =
(15) 5 x 11 =

(16) 8 x 10 =
(17) 12 x 2 =
(18) 9 x 10 =

(19) 12 x 2 =
(20) 10 x 7 =
(21) 3 x 1 =

(22) 1 x 8 =
(23) 8 x 11 =
(24) 9 x 11 =

(25) 4 x 2 =
(26) 12 x 4 =
(27) 8 x 6 =

(28) 7 x 5 =
(29) 4 x 4 =
(30) 6 x 10 =

(31) 9 x 3 =
(32) 2 x 6 =
(33) 2 x 4 =

(34) 7 x 1 =
(35) 2 x 6 =
(36) 9 x 4 =

(37) 12 x 2 =
(38) 7 x 11 =
(39) 4 x 3 =

(40) 2 x 4 =
(41) 8 x 3 =
(42) 10 x 8 =

(43) 8 x 4 =
(44) 8 x 1 =
(45) 8 x 10 =

(46) 1 x 6 =
(47) 1 x 7 =
(48) 7 x 4 =

(49) 3 x 10 =
(50) 1 x 5 =
(51) 3 x 2 =

(52) 1 x 7 =
(53) 9 x 6 =
(54) 10 x 4 =

(55) 4 x 4 =
(56) 3 x 6 =
(57) 11 x 5 =

(58) 5 x 12 =
(59) 10 x 2 =
(60) 10 x 3 =

(1) 10 x 12 =   (2) 5 x 8 =   (3) 4 x 8 =

(4) 9 x 12 =   (5) 1 x 1 =   (6) 12 x 9 =

(7) 12 x 10 =   (8) 5 x 2 =   (9) 3 x 10 =

(10) 9 x 1 =   (11) 9 x 5 =   (12) 3 x 6 =

(13) 1 x 3 =   (14) 1 x 10 =   (15) 2 x 8 =

(16) 8 x 3 =   (17) 1 x 3 =   (18) 3 x 9 =

(19) 1 x 6 =   (20) 9 x 9 =   (21) 1 x 11 =

(22) 4 x 11 =   (23) 10 x 3 =   (24) 1 x 7 =

(25) 2 x 11 =   (26) 11 x 9 =   (27) 8 x 8 =

(28) 7 x 12 =   (29) 8 x 8 =   (30) 4 x 5 =

(31) 12 x 3 =   (32) 8 x 1 =   (33) 7 x 1 =

(34) 8 x 2 =   (35) 3 x 3 =   (36) 1 x 2 =

(37) 6 x 9 =   (38) 5 x 3 =   (39) 4 x 5 =

(40) 9 x 7 =   (41) 9 x 6 =   (42) 7 x 4 =

(43) 2 x 9 =   (44) 9 x 8 =   (45) 9 x 4 =

(46) 10 x 5 =   (47) 9 x 6 =   (48) 4 x 6 =

(49) 12 x 11 =   (50) 5 x 10 =   (51) 6 x 7 =

(52) 6 x 2 =   (53) 8 x 7 =   (54) 4 x 5 =

(55) 8 x 9 =   (56) 7 x 5 =   (57) 2 x 3 =

(58) 11 x 2 =   (59) 2 x 4 =   (60) 10 x 2 =

(1) 9 x 9 =

(2) 7 x 4 =

(3) 12 x 10 =

(4) 6 x 5 =

(5) 9 x 5 =

(6) 6 x 3 =

(7) 7 x 9 =

(8) 4 x 6 =

(9) 10 x 2 =

(10) 1 x 8 =

(11) 1 x 7 =

(12) 9 x 11 =

(13) 1 x 9 =

(14) 3 x 11 =

(15) 3 x 10 =

(16) 9 x 3 =

(17) 3 x 3 =

(18) 4 x 9 =

(19) 5 x 10 =

(20) 2 x 3 =

(21) 6 x 1 =

(22) 6 x 12 =

(23) 8 x 9 =

(24) 3 x 3 =

(25) 8 x 7 =

(26) 7 x 3 =

(27) 2 x 4 =

(28) 5 x 1 =

(29) 5 x 11 =

(30) 3 x 9 =

(31) 3 x 7 =

(32) 12 x 2 =

(33) 4 x 5 =

(34) 2 x 10 =

(35) 1 x 10 =

(36) 10 x 6 =

(37) 10 x 5 =

(38) 10 x 2 =

(39) 11 x 1 =

(40) 3 x 6 =

(41) 9 x 11 =

(42) 12 x 1 =

(43) 1 x 9 =

(44) 2 x 10 =

(45) 9 x 10 =

(46) 1 x 8 =

(47) 8 x 11 =

(48) 12 x 2 =

(49) 3 x 4 =

(50) 8 x 8 =

(51) 3 x 2 =

(52) 10 x 11 =

(53) 4 x 2 =

(54) 1 x 11 =

(55) 6 x 6 =

(56) 10 x 7 =

(57) 10 x 11 =

(58) 7 x 9 =

(59) 5 x 1 =

(60) 3 x 12 =

| Day: | 38 | Date: | | Score: | /60 |
| Name: | | Time: | : | Rating: | ☆☆☆☆☆ |

(1) 7 x 2 =

(2) 11 x 2 =

(3) 6 x 1 =

(4) 5 x 4 =

(5) 10 x 5 =

(6) 3 x 9 =

(7) 5 x 2 =

(8) 9 x 6 =

(9) 2 x 1 =

(10) 7 x 7 =

(11) 1 x 2 =

(12) 6 x 9 =

(13) 8 x 9 =

(14) 8 x 11 =

(15) 2 x 2 =

(16) 10 x 11 =

(17) 6 x 10 =

(18) 3 x 6 =

(19) 9 x 9 =

(20) 5 x 9 =

(21) 2 x 1 =

(22) 1 x 12 =

(23) 10 x 1 =

(24) 5 x 2 =

(25) 3 x 2 =

(26) 9 x 10 =

(27) 10 x 3 =

(28) 9 x 4 =

(29) 6 x 8 =

(30) 12 x 10 =

(31) 7 x 7 =

(32) 10 x 8 =

(33) 7 x 10 =

(34) 2 x 1 =

(35) 1 x 9 =

(36) 7 x 7 =

(37) 10 x 4 =

(38) 4 x 9 =

(39) 4 x 12 =

(40) 11 x 6 =

(41) 6 x 6 =

(42) 11 x 1 =

(43) 3 x 12 =

(44) 7 x 5 =

(45) 3 x 8 =

(46) 2 x 3 =

(47) 7 x 8 =

(48) 7 x 6 =

(49) 2 x 7 =

(50) 2 x 11 =

(51) 1 x 5 =

(52) 11 x 5 =

(53) 10 x 10 =

(54) 3 x 12 =

(55) 11 x 3 =

(56) 7 x 3 =

(57) 5 x 8 =

(58) 5 x 9 =

(59) 9 x 4 =

(60) 4 x 9 =

(1) 7 x 3 =

(2) 6 x 12 =

(3) 7 x 4 =

(4) 7 x 11 =

(5) 3 x 8 =

(6) 12 x 1 =

(7) 9 x 8 =

(8) 10 x 12 =

(9) 6 x 8 =

(10) 6 x 11 =

(11) 1 x 9 =

(12) 11 x 4 =

(13) 1 x 1 =

(14) 9 x 3 =

(15) 8 x 12 =

(16) 9 x 6 =

(17) 12 x 10 =

(18) 4 x 5 =

(19) 10 x 2 =

(20) 3 x 9 =

(21) 8 x 12 =

(22) 9 x 12 =

(23) 2 x 8 =

(24) 4 x 12 =

(25) 6 x 3 =

(26) 10 x 5 =

(27) 12 x 3 =

(28) 4 x 11 =

(29) 6 x 11 =

(30) 2 x 7 =

(31) 4 x 2 =

(32) 3 x 12 =

(33) 7 x 2 =

(34) 1 x 12 =

(35) 6 x 2 =

(36) 12 x 8 =

(37) 2 x 1 =

(38) 3 x 7 =

(39) 6 x 11 =

(40) 4 x 10 =

(41) 8 x 9 =

(42) 2 x 9 =

(43) 6 x 7 =

(44) 4 x 7 =

(45) 12 x 12 =

(46) 10 x 2 =

(47) 5 x 7 =

(48) 10 x 10 =

(49) 5 x 12 =

(50) 10 x 5 =

(51) 11 x 4 =

(52) 9 x 12 =

(53) 5 x 3 =

(54) 12 x 10 =

(55) 12 x 3 =

(56) 9 x 4 =

(57) 10 x 10 =

(58) 11 x 8 =

(59) 7 x 6 =

(60) 5 x 5 =

(1) 7 x 11 =

(2) 6 x 6 =

(3) 11 x 7 =

(4) 9 x 5 =

(5) 6 x 8 =

(6) 11 x 1 =

(7) 9 x 2 =

(8) 9 x 4 =

(9) 4 x 10 =

(10) 4 x 6 =

(11) 1 x 12 =

(12) 12 x 9 =

(13) 1 x 11 =

(14) 5 x 12 =

(15) 3 x 9 =

(16) 10 x 4 =

(17) 11 x 2 =

(18) 4 x 1 =

(19) 3 x 9 =

(20) 3 x 5 =

(21) 6 x 11 =

(22) 4 x 3 =

(23) 12 x 1 =

(24) 6 x 6 =

(25) 7 x 12 =

(26) 3 x 1 =

(27) 12 x 6 =

(28) 11 x 9 =

(29) 4 x 1 =

(30) 5 x 4 =

(31) 7 x 2 =

(32) 10 x 9 =

(33) 8 x 2 =

(34) 3 x 3 =

(35) 11 x 12 =

(36) 2 x 8 =

(37) 7 x 5 =

(38) 8 x 11 =

(39) 10 x 11 =

(40) 3 x 8 =

(41) 7 x 2 =

(42) 4 x 2 =

(43) 1 x 7 =

(44) 11 x 5 =

(45) 10 x 11 =

(46) 5 x 8 =

(47) 5 x 3 =

(48) 3 x 12 =

(49) 7 x 4 =

(50) 8 x 4 =

(51) 8 x 8 =

(52) 12 x 12 =

(53) 5 x 2 =

(54) 6 x 6 =

(55) 4 x 5 =

(56) 7 x 2 =

(57) 5 x 12 =

(58) 6 x 3 =

(59) 4 x 3 =

(60) 11 x 6 =

(1) 7 x 5 =

(2) 8 x 10 =

(3) 3 x 11 =

(4) 10 x 9 =

(5) 3 x 5 =

(6) 10 x 7 =

(7) 9 x 11 =

(8) 7 x 12 =

(9) 9 x 12 =

(10) 3 x 6 =

(11) 2 x 2 =

(12) 12 x 2 =

(13) 10 x 10 =

(14) 3 x 1 =

(15) 5 x 11 =

(16) 6 x 3 =

(17) 2 x 10 =

(18) 7 x 10 =

(19) 2 x 1 =

(20) 11 x 9 =

(21) 8 x 10 =

(22) 1 x 9 =

(23) 5 x 12 =

(24) 11 x 10 =

(25) 5 x 10 =

(26) 4 x 11 =

(27) 7 x 2 =

(28) 12 x 2 =

(29) 8 x 12 =

(30) 1 x 1 =

(31) 11 x 8 =

(32) 3 x 3 =

(33) 4 x 3 =

(34) 3 x 10 =

(35) 9 x 10 =

(36) 6 x 7 =

(37) 7 x 10 =

(38) 3 x 4 =

(39) 2 x 11 =

(40) 10 x 4 =

(41) 4 x 7 =

(42) 6 x 7 =

(43) 5 x 4 =

(44) 9 x 2 =

(45) 9 x 9 =

(46) 8 x 7 =

(47) 5 x 7 =

(48) 6 x 4 =

(49) 12 x 1 =

(50) 10 x 2 =

(51) 6 x 12 =

(52) 9 x 5 =

(53) 4 x 2 =

(54) 10 x 6 =

(55) 11 x 12 =

(56) 10 x 10 =

(57) 7 x 9 =

(58) 5 x 8 =

(59) 5 x 10 =

(60) 9 x 9 =

Day:	42	Date:		Score:	/60
Name:		Time:	:	Rating:	☆☆☆☆☆

(1) 4 x 7 =

(2) 5 x 11 =

(3) 8 x 2 =

(4) 2 x 8 =

(5) 12 x 4 =

(6) 12 x 5 =

(7) 10 x 2 =

(8) 11 x 10 =

(9) 7 x 7 =

(10) 3 x 7 =

(11) 7 x 3 =

(12) 8 x 12 =

(13) 8 x 5 =

(14) 8 x 4 =

(15) 6 x 12 =

(16) 12 x 11 =

(17) 1 x 1 =

(18) 12 x 4 =

(19) 1 x 11 =

(20) 8 x 2 =

(21) 8 x 7 =

(22) 10 x 5 =

(23) 2 x 10 =

(24) 12 x 12 =

(25) 11 x 3 =

(26) 5 x 7 =

(27) 11 x 12 =

(28) 4 x 12 =

(29) 10 x 5 =

(30) 10 x 8 =

(31) 4 x 7 =

(32) 4 x 8 =

(33) 7 x 12 =

(34) 5 x 11 =

(35) 7 x 10 =

(36) 8 x 6 =

(37) 10 x 10 =

(38) 7 x 7 =

(39) 7 x 11 =

(40) 2 x 6 =

(41) 7 x 8 =

(42) 12 x 4 =

(43) 10 x 11 =

(44) 9 x 3 =

(45) 5 x 12 =

(46) 4 x 10 =

(47) 1 x 11 =

(48) 6 x 2 =

(49) 3 x 12 =

(50) 8 x 10 =

(51) 3 x 6 =

(52) 2 x 1 =

(53) 8 x 8 =

(54) 12 x 8 =

(55) 3 x 10 =

(56) 12 x 2 =

(57) 6 x 8 =

(58) 8 x 11 =

(59) 11 x 4 =

(60) 6 x 2 =

(1) 8 x 8 =
(2) 10 x 9 =
(3) 10 x 3 =

(4) 1 x 7 =
(5) 6 x 3 =
(6) 4 x 8 =

(7) 2 x 1 =
(8) 9 x 9 =
(9) 3 x 11 =

(10) 7 x 6 =
(11) 9 x 12 =
(12) 10 x 9 =

(13) 12 x 5 =
(14) 1 x 1 =
(15) 5 x 7 =

(16) 5 x 6 =
(17) 5 x 3 =
(18) 2 x 3 =

(19) 9 x 10 =
(20) 8 x 12 =
(21) 5 x 2 =

(22) 8 x 6 =
(23) 4 x 5 =
(24) 11 x 12 =

(25) 9 x 7 =
(26) 11 x 5 =
(27) 7 x 11 =

(28) 4 x 1 =
(29) 6 x 11 =
(30) 3 x 9 =

(31) 6 x 12 =
(32) 5 x 10 =
(33) 5 x 12 =

(34) 7 x 5 =
(35) 4 x 11 =
(36) 9 x 1 =

(37) 2 x 4 =
(38) 11 x 8 =
(39) 11 x 4 =

(40) 12 x 7 =
(41) 9 x 9 =
(42) 4 x 5 =

(43) 9 x 1 =
(44) 10 x 7 =
(45) 2 x 1 =

(46) 5 x 2 =
(47) 2 x 1 =
(48) 4 x 7 =

(49) 1 x 2 =
(50) 2 x 12 =
(51) 1 x 12 =

(52) 8 x 5 =
(53) 1 x 11 =
(54) 6 x 12 =

(55) 1 x 9 =
(56) 11 x 7 =
(57) 10 x 8 =

(58) 8 x 1 =
(59) 4 x 11 =
(60) 1 x 12 =

(1) 11 x 6 =   (2) 7 x 3 =   (3) 3 x 1 =

(4) 3 x 11 =   (5) 8 x 8 =   (6) 9 x 9 =

(7) 1 x 6 =   (8) 8 x 5 =   (9) 2 x 2 =

(10) 4 x 11 =   (11) 8 x 8 =   (12) 5 x 9 =

(13) 6 x 7 =   (14) 8 x 7 =   (15) 10 x 7 =

(16) 4 x 8 =   (17) 5 x 1 =   (18) 10 x 9 =

(19) 8 x 7 =   (20) 1 x 8 =   (21) 7 x 4 =

(22) 12 x 2 =   (23) 11 x 5 =   (24) 11 x 4 =

(25) 1 x 2 =   (26) 6 x 12 =   (27) 10 x 1 =

(28) 2 x 6 =   (29) 1 x 5 =   (30) 8 x 9 =

(31) 5 x 3 =   (32) 7 x 5 =   (33) 3 x 5 =

(34) 8 x 11 =   (35) 2 x 12 =   (36) 8 x 1 =

(37) 8 x 4 =   (38) 4 x 10 =   (39) 12 x 8 =

(40) 5 x 11 =   (41) 3 x 2 =   (42) 1 x 1 =

(43) 11 x 2 =   (44) 11 x 1 =   (45) 1 x 10 =

(46) 9 x 4 =   (47) 9 x 11 =   (48) 9 x 1 =

(49) 6 x 2 =   (50) 2 x 6 =   (51) 5 x 10 =

(52) 1 x 4 =   (53) 9 x 7 =   (54) 4 x 2 =

(55) 9 x 8 =   (56) 2 x 10 =   (57) 12 x 2 =

(58) 7 x 9 =   (59) 6 x 1 =   (60) 2 x 2 =

(1) $6 \times 10 =$

(2) $9 \times 12 =$

(3) $8 \times 12 =$

(4) $8 \times 7 =$

(5) $6 \times 8 =$

(6) $3 \times 12 =$

(7) $12 \times 10 =$

(8) $4 \times 6 =$

(9) $1 \times 9 =$

(10) $2 \times 1 =$

(11) $10 \times 6 =$

(12) $4 \times 5 =$

(13) $9 \times 3 =$

(14) $11 \times 8 =$

(15) $8 \times 8 =$

(16) $7 \times 10 =$

(17) $4 \times 2 =$

(18) $6 \times 8 =$

(19) $12 \times 2 =$

(20) $4 \times 9 =$

(21) $11 \times 11 =$

(22) $2 \times 5 =$

(23) $5 \times 6 =$

(24) $1 \times 2 =$

(25) $1 \times 2 =$

(26) $7 \times 2 =$

(27) $9 \times 2 =$

(28) $2 \times 10 =$

(29) $6 \times 5 =$

(30) $11 \times 9 =$

(31) $2 \times 4 =$

(32) $9 \times 5 =$

(33) $2 \times 2 =$

(34) $1 \times 9 =$

(35) $6 \times 12 =$

(36) $4 \times 3 =$

(37) $5 \times 3 =$

(38) $10 \times 9 =$

(39) $6 \times 7 =$

(40) $6 \times 6 =$

(41) $1 \times 7 =$

(42) $1 \times 4 =$

(43) $6 \times 7 =$

(44) $11 \times 12 =$

(45) $4 \times 4 =$

(46) $1 \times 7 =$

(47) $11 \times 2 =$

(48) $1 \times 9 =$

(49) $10 \times 7 =$

(50) $11 \times 12 =$

(51) $8 \times 9 =$

(52) $2 \times 3 =$

(53) $1 \times 3 =$

(54) $8 \times 10 =$

(55) $3 \times 6 =$

(56) $2 \times 4 =$

(57) $1 \times 4 =$

(58) $2 \times 5 =$

(59) $7 \times 12 =$

(60) $8 \times 8 =$

(1) 5 x 11 =

(2) 10 x 2 =

(3) 4 x 5 =

(4) 10 x 10 =

(5) 1 x 10 =

(6) 4 x 11 =

(7) 9 x 8 =

(8) 3 x 9 =

(9) 7 x 8 =

(10) 12 x 11 =

(11) 3 x 5 =

(12) 8 x 12 =

(13) 3 x 8 =

(14) 8 x 3 =

(15) 12 x 11 =

(16) 8 x 12 =

(17) 3 x 8 =

(18) 11 x 12 =

(19) 1 x 11 =

(20) 2 x 12 =

(21) 4 x 5 =

(22) 7 x 1 =

(23) 11 x 8 =

(24) 4 x 8 =

(25) 4 x 9 =

(26) 8 x 6 =

(27) 5 x 4 =

(28) 10 x 3 =

(29) 4 x 9 =

(30) 4 x 4 =

(31) 11 x 7 =

(32) 11 x 3 =

(33) 2 x 9 =

(34) 11 x 8 =

(35) 9 x 10 =

(36) 7 x 9 =

(37) 2 x 5 =

(38) 3 x 12 =

(39) 12 x 2 =

(40) 7 x 7 =

(41) 9 x 12 =

(42) 1 x 11 =

(43) 3 x 9 =

(44) 8 x 1 =

(45) 12 x 6 =

(46) 5 x 7 =

(47) 2 x 12 =

(48) 3 x 6 =

(49) 6 x 7 =

(50) 7 x 3 =

(51) 5 x 1 =

(52) 4 x 12 =

(53) 5 x 9 =

(54) 3 x 6 =

(55) 11 x 6 =

(56) 5 x 4 =

(57) 8 x 9 =

(58) 10 x 7 =

(59) 10 x 12 =

(60) 4 x 2 =

| Day: | 47 | Date: | | Score: | /60 |
| Name: | | Time: | : | Rating: | ☆☆☆☆☆ |

(1) 12 x 6 =

(2) 6 x 6 =

(3) 1 x 3 =

(4) 1 x 11 =

(5) 12 x 8 =

(6) 5 x 5 =

(7) 3 x 12 =

(8) 11 x 8 =

(9) 12 x 11 =

(10) 2 x 10 =

(11) 7 x 1 =

(12) 3 x 12 =

(13) 7 x 1 =

(14) 10 x 9 =

(15) 11 x 10 =

(16) 3 x 10 =

(17) 7 x 6 =

(18) 12 x 4 =

(19) 1 x 4 =

(20) 6 x 11 =

(21) 2 x 12 =

(22) 8 x 9 =

(23) 12 x 10 =

(24) 2 x 2 =

(25) 10 x 3 =

(26) 2 x 11 =

(27) 7 x 7 =

(28) 2 x 10 =

(29) 11 x 4 =

(30) 3 x 10 =

(31) 9 x 8 =

(32) 12 x 3 =

(33) 8 x 4 =

(34) 2 x 11 =

(35) 5 x 3 =

(36) 10 x 8 =

(37) 3 x 8 =

(38) 4 x 7 =

(39) 6 x 5 =

(40) 5 x 6 =

(41) 5 x 2 =

(42) 7 x 4 =

(43) 4 x 1 =

(44) 7 x 4 =

(45) 2 x 10 =

(46) 4 x 8 =

(47) 10 x 5 =

(48) 5 x 4 =

(49) 6 x 6 =

(50) 2 x 9 =

(51) 9 x 10 =

(52) 1 x 2 =

(53) 10 x 5 =

(54) 8 x 7 =

(55) 12 x 7 =

(56) 10 x 4 =

(57) 7 x 4 =

(58) 11 x 12 =

(59) 12 x 10 =

(60) 2 x 7 =

(1) 5 x 4 =

(2) 11 x 2 =

(3) 9 x 4 =

(4) 11 x 7 =

(5) 12 x 12 =

(6) 9 x 4 =

(7) 7 x 3 =

(8) 8 x 8 =

(9) 11 x 2 =

(10) 8 x 4 =

(11) 6 x 4 =

(12) 10 x 1 =

(13) 12 x 12 =

(14) 8 x 7 =

(15) 12 x 6 =

(16) 4 x 10 =

(17) 11 x 9 =

(18) 6 x 12 =

(19) 4 x 12 =

(20) 11 x 8 =

(21) 8 x 8 =

(22) 12 x 5 =

(23) 5 x 8 =

(24) 6 x 7 =

(25) 2 x 12 =

(26) 6 x 3 =

(27) 6 x 2 =

(28) 3 x 4 =

(29) 11 x 6 =

(30) 5 x 6 =

(31) 9 x 9 =

(32) 6 x 8 =

(33) 8 x 7 =

(34) 10 x 3 =

(35) 4 x 12 =

(36) 12 x 1 =

(37) 11 x 4 =

(38) 7 x 3 =

(39) 1 x 7 =

(40) 12 x 10 =

(41) 3 x 9 =

(42) 10 x 9 =

(43) 2 x 3 =

(44) 10 x 11 =

(45) 7 x 7 =

(46) 4 x 10 =

(47) 1 x 8 =

(48) 7 x 5 =

(49) 3 x 7 =

(50) 7 x 5 =

(51) 1 x 7 =

(52) 2 x 5 =

(53) 3 x 10 =

(54) 4 x 4 =

(55) 12 x 2 =

(56) 5 x 7 =

(57) 2 x 2 =

(58) 12 x 5 =

(59) 12 x 9 =

(60) 9 x 1 =

(1) $8 \times 5 =$

(2) $7 \times 10 =$

(3) $2 \times 12 =$

(4) $10 \times 7 =$

(5) $7 \times 3 =$

(6) $3 \times 2 =$

(7) $11 \times 4 =$

(8) $7 \times 3 =$

(9) $2 \times 6 =$

(10) $9 \times 4 =$

(11) $11 \times 9 =$

(12) $7 \times 5 =$

(13) $2 \times 12 =$

(14) $1 \times 8 =$

(15) $6 \times 5 =$

(16) $8 \times 12 =$

(17) $8 \times 10 =$

(18) $12 \times 1 =$

(19) $3 \times 8 =$

(20) $10 \times 9 =$

(21) $5 \times 6 =$

(22) $12 \times 1 =$

(23) $11 \times 12 =$

(24) $8 \times 6 =$

(25) $10 \times 7 =$

(26) $9 \times 4 =$

(27) $11 \times 10 =$

(28) $4 \times 5 =$

(29) $6 \times 5 =$

(30) $6 \times 10 =$

(31) $10 \times 6 =$

(32) $4 \times 1 =$

(33) $4 \times 10 =$

(34) $9 \times 6 =$

(35) $10 \times 6 =$

(36) $6 \times 1 =$

(37) $3 \times 5 =$

(38) $7 \times 8 =$

(39) $8 \times 1 =$

(40) $6 \times 3 =$

(41) $5 \times 10 =$

(42) $9 \times 12 =$

(43) $1 \times 7 =$

(44) $9 \times 10 =$

(45) $6 \times 1 =$

(46) $6 \times 11 =$

(47) $12 \times 3 =$

(48) $6 \times 4 =$

(49) $9 \times 8 =$

(50) $12 \times 5 =$

(51) $10 \times 6 =$

(52) $5 \times 1 =$

(53) $7 \times 2 =$

(54) $4 \times 3 =$

(55) $11 \times 10 =$

(56) $5 \times 8 =$

(57) $11 \times 5 =$

(58) $8 \times 7 =$

(59) $6 \times 5 =$

(60) $6 \times 11 =$

(1) $6 \times 11 =$

(2) $12 \times 10 =$

(3) $5 \times 2 =$

(4) $8 \times 11 =$

(5) $3 \times 11 =$

(6) $6 \times 1 =$

(7) $9 \times 9 =$

(8) $8 \times 10 =$

(9) $6 \times 8 =$

(10) $3 \times 10 =$

(11) $9 \times 8 =$

(12) $1 \times 6 =$

(13) $2 \times 10 =$

(14) $10 \times 10 =$

(15) $12 \times 8 =$

(16) $5 \times 5 =$

(17) $5 \times 7 =$

(18) $12 \times 2 =$

(19) $4 \times 11 =$

(20) $7 \times 5 =$

(21) $2 \times 2 =$

(22) $6 \times 1 =$

(23) $5 \times 3 =$

(24) $9 \times 7 =$

(25) $11 \times 6 =$

(26) $1 \times 10 =$

(27) $10 \times 10 =$

(28) $12 \times 12 =$

(29) $7 \times 10 =$

(30) $4 \times 9 =$

(31) $10 \times 4 =$

(32) $3 \times 7 =$

(33) $5 \times 1 =$

(34) $7 \times 8 =$

(35) $10 \times 11 =$

(36) $10 \times 1 =$

(37) $11 \times 7 =$

(38) $9 \times 11 =$

(39) $8 \times 6 =$

(40) $1 \times 11 =$

(41) $1 \times 2 =$

(42) $11 \times 12 =$

(43) $4 \times 9 =$

(44) $8 \times 5 =$

(45) $5 \times 8 =$

(46) $4 \times 9 =$

(47) $5 \times 10 =$

(48) $8 \times 6 =$

(49) $7 \times 6 =$

(50) $11 \times 3 =$

(51) $1 \times 8 =$

(52) $7 \times 1 =$

(53) $4 \times 11 =$

(54) $10 \times 7 =$

(55) $1 \times 7 =$

(56) $9 \times 10 =$

(57) $11 \times 3 =$

(58) $8 \times 9 =$

(59) $10 \times 3 =$

(60) $7 \times 8 =$

(1) $2 \div 2 =$

(2) $10 \div 5 =$

(3) $72 \div 12 =$

(4) $10 \div 10 =$

(5) $88 \div 11 =$

(6) $32 \div 4 =$

(7) $20 \div 10 =$

(8) $63 \div 7 =$

(9) $16 \div 2 =$

(10) $108 \div 9 =$

(11) $21 \div 3 =$

(12) $32 \div 8 =$

(13) $20 \div 10 =$

(14) $24 \div 8 =$

(15) $4 \div 1 =$

(16) $44 \div 11 =$

(17) $48 \div 8 =$

(18) $88 \div 11 =$

(19) $33 \div 11 =$

(20) $36 \div 12 =$

(21) $8 \div 4 =$

(22) $70 \div 10 =$

(23) $5 \div 1 =$

(24) $60 \div 10 =$

(25) $54 \div 9 =$

(26) $63 \div 9 =$

(27) $80 \div 8 =$

(28) $21 \div 7 =$

(29) $15 \div 3 =$

(30) $2 \div 2 =$

(31) $11 \div 11 =$

(32) $50 \div 5 =$

(33) $7 \div 7 =$

(34) $132 \div 12 =$

(35) $54 \div 6 =$

(36) $10 \div 10 =$

(37) $50 \div 5 =$

(38) $18 \div 2 =$

(39) $8 \div 1 =$

(40) $72 \div 6 =$

(41) $20 \div 10 =$

(42) $90 \div 10 =$

(43) $42 \div 6 =$

(44) $16 \div 2 =$

(45) $7 \div 7 =$

(46) $48 \div 4 =$

(47) $66 \div 6 =$

(48) $36 \div 4 =$

(49) $44 \div 4 =$

(50) $36 \div 3 =$

(51) $10 \div 2 =$

(52) $32 \div 8 =$

(53) $12 \div 12 =$

(54) $20 \div 5 =$

(55) $5 \div 1 =$

(56) $20 \div 2 =$

(57) $33 \div 3 =$

(58) $3 \div 3 =$

(59) $55 \div 5 =$

(60) $56 \div 8 =$

(1) 30 ÷ 10 =

(2) 36 ÷ 6 =

(3) 30 ÷ 3 =

(4) 42 ÷ 7 =

(5) 55 ÷ 11 =

(6) 49 ÷ 7 =

(7) 12 ÷ 2 =

(8) 42 ÷ 7 =

(9) 81 ÷ 9 =

(10) 4 ÷ 1 =

(11) 20 ÷ 5 =

(12) 18 ÷ 6 =

(13) 99 ÷ 9 =

(14) 15 ÷ 3 =

(15) 6 ÷ 6 =

(16) 121 ÷ 11 =

(17) 36 ÷ 12 =

(18) 8 ÷ 2 =

(19) 10 ÷ 1 =

(20) 49 ÷ 7 =

(21) 16 ÷ 2 =

(22) 14 ÷ 2 =

(23) 63 ÷ 7 =

(24) 6 ÷ 3 =

(25) 2 ÷ 2 =

(26) 36 ÷ 6 =

(27) 3 ÷ 1 =

(28) 60 ÷ 10 =

(29) 50 ÷ 5 =

(30) 14 ÷ 7 =

(31) 48 ÷ 4 =

(32) 84 ÷ 7 =

(33) 63 ÷ 9 =

(34) 56 ÷ 8 =

(35) 64 ÷ 8 =

(36) 24 ÷ 2 =

(37) 121 ÷ 11 =

(38) 24 ÷ 2 =

(39) 35 ÷ 5 =

(40) 50 ÷ 5 =

(41) 120 ÷ 10 =

(42) 88 ÷ 8 =

(43) 60 ÷ 10 =

(44) 6 ÷ 1 =

(45) 99 ÷ 9 =

(46) 15 ÷ 3 =

(47) 10 ÷ 2 =

(48) 8 ÷ 8 =

(49) 8 ÷ 2 =

(50) 20 ÷ 2 =

(51) 60 ÷ 12 =

(52) 12 ÷ 2 =

(53) 11 ÷ 1 =

(54) 15 ÷ 5 =

(55) 70 ÷ 10 =

(56) 45 ÷ 5 =

(57) 12 ÷ 2 =

(58) 20 ÷ 4 =

(59) 30 ÷ 5 =

(60) 9 ÷ 9 =

(1) $24 \div 3 =$

(2) $6 \div 1 =$

(3) $4 \div 2 =$

(4) $110 \div 11 =$

(5) $56 \div 7 =$

(6) $96 \div 12 =$

(7) $35 \div 7 =$

(8) $7 \div 7 =$

(9) $49 \div 7 =$

(10) $120 \div 12 =$

(11) $56 \div 7 =$

(12) $108 \div 12 =$

(13) $77 \div 7 =$

(14) $66 \div 6 =$

(15) $48 \div 6 =$

(16) $12 \div 3 =$

(17) $100 \div 10 =$

(18) $9 \div 1 =$

(19) $144 \div 12 =$

(20) $48 \div 8 =$

(21) $48 \div 4 =$

(22) $72 \div 9 =$

(23) $108 \div 12 =$

(24) $63 \div 9 =$

(25) $6 \div 6 =$

(26) $27 \div 9 =$

(27) $30 \div 10 =$

(28) $8 \div 8 =$

(29) $48 \div 6 =$

(30) $100 \div 10 =$

(31) $24 \div 6 =$

(32) $70 \div 10 =$

(33) $55 \div 11 =$

(34) $21 \div 7 =$

(35) $28 \div 4 =$

(36) $7 \div 1 =$

(37) $72 \div 12 =$

(38) $66 \div 11 =$

(39) $10 \div 10 =$

(40) $14 \div 2 =$

(41) $54 \div 9 =$

(42) $2 \div 1 =$

(43) $40 \div 10 =$

(44) $72 \div 9 =$

(45) $4 \div 2 =$

(46) $96 \div 8 =$

(47) $8 \div 1 =$

(48) $72 \div 8 =$

(49) $45 \div 5 =$

(50) $132 \div 11 =$

(51) $63 \div 7 =$

(52) $2 \div 1 =$

(53) $48 \div 12 =$

(54) $24 \div 2 =$

(55) $18 \div 3 =$

(56) $32 \div 8 =$

(57) $70 \div 7 =$

(58) $60 \div 6 =$

(59) $10 \div 2 =$

(60) $12 \div 3 =$

(1)　$96 \div 8 =$

(2)　$33 \div 11 =$

(3)　$22 \div 2 =$

(4)　$2 \div 2 =$

(5)　$18 \div 3 =$

(6)　$44 \div 4 =$

(7)　$24 \div 4 =$

(8)　$10 \div 2 =$

(9)　$80 \div 10 =$

(10)　$99 \div 11 =$

(11)　$15 \div 5 =$

(12)　$120 \div 10 =$

(13)　$56 \div 8 =$

(14)　$22 \div 11 =$

(15)　$18 \div 6 =$

(16)　$21 \div 3 =$

(17)　$30 \div 3 =$

(18)　$144 \div 12 =$

(19)　$9 \div 3 =$

(20)　$6 \div 3 =$

(21)　$36 \div 12 =$

(22)　$48 \div 12 =$

(23)　$4 \div 1 =$

(24)　$10 \div 1 =$

(25)　$66 \div 6 =$

(26)　$80 \div 10 =$

(27)　$50 \div 5 =$

(28)　$60 \div 6 =$

(29)　$6 \div 3 =$

(30)　$48 \div 6 =$

(31)　$42 \div 6 =$

(32)　$11 \div 11 =$

(33)　$24 \div 12 =$

(34)　$7 \div 1 =$

(35)　$30 \div 3 =$

(36)　$14 \div 7 =$

(37)　$20 \div 2 =$

(38)　$90 \div 9 =$

(39)　$42 \div 6 =$

(40)　$35 \div 7 =$

(41)　$44 \div 11 =$

(42)　$63 \div 7 =$

(43)　$84 \div 7 =$

(44)　$15 \div 5 =$

(45)　$96 \div 8 =$

(46)　$45 \div 9 =$

(47)　$18 \div 6 =$

(48)　$50 \div 10 =$

(49)　$16 \div 4 =$

(50)　$77 \div 7 =$

(51)　$18 \div 9 =$

(52)　$35 \div 7 =$

(53)　$22 \div 2 =$

(54)　$16 \div 8 =$

(55)　$90 \div 10 =$

(56)　$10 \div 10 =$

(57)　$35 \div 7 =$

(58)　$48 \div 4 =$

(59)　$66 \div 11 =$

(60)　$88 \div 11 =$

(1) 24 ÷ 6 =

(2) 66 ÷ 6 =

(3) 35 ÷ 7 =

(4) 88 ÷ 11 =

(5) 20 ÷ 2 =

(6) 72 ÷ 6 =

(7) 55 ÷ 5 =

(8) 8 ÷ 8 =

(9) 27 ÷ 3 =

(10) 24 ÷ 4 =

(11) 120 ÷ 10 =

(12) 18 ÷ 9 =

(13) 120 ÷ 10 =

(14) 35 ÷ 5 =

(15) 90 ÷ 9 =

(16) 84 ÷ 7 =

(17) 3 ÷ 1 =

(18) 84 ÷ 12 =

(19) 21 ÷ 7 =

(20) 36 ÷ 9 =

(21) 28 ÷ 4 =

(22) 60 ÷ 5 =

(23) 132 ÷ 11 =

(24) 81 ÷ 9 =

(25) 9 ÷ 9 =

(26) 8 ÷ 1 =

(27) 90 ÷ 9 =

(28) 70 ÷ 7 =

(29) 6 ÷ 1 =

(30) 9 ÷ 9 =

(31) 66 ÷ 6 =

(32) 24 ÷ 8 =

(33) 8 ÷ 8 =

(34) 36 ÷ 3 =

(35) 70 ÷ 7 =

(36) 30 ÷ 6 =

(37) 12 ÷ 12 =

(38) 8 ÷ 8 =

(39) 60 ÷ 10 =

(40) 84 ÷ 7 =

(41) 9 ÷ 3 =

(42) 16 ÷ 8 =

(43) 48 ÷ 8 =

(44) 72 ÷ 9 =

(45) 15 ÷ 3 =

(46) 50 ÷ 10 =

(47) 22 ÷ 11 =

(48) 60 ÷ 6 =

(49) 33 ÷ 11 =

(50) 21 ÷ 3 =

(51) 12 ÷ 4 =

(52) 70 ÷ 10 =

(53) 56 ÷ 8 =

(54) 50 ÷ 10 =

(55) 60 ÷ 10 =

(56) 8 ÷ 8 =

(57) 45 ÷ 5 =

(58) 24 ÷ 3 =

(59) 12 ÷ 2 =

(60) 12 ÷ 12 =

(1) $77 \div 11 =$

(2) $28 \div 7 =$

(3) $2 \div 1 =$

(4) $32 \div 8 =$

(5) $11 \div 11 =$

(6) $60 \div 12 =$

(7) $40 \div 8 =$

(8) $70 \div 7 =$

(9) $5 \div 1 =$

(10) $42 \div 6 =$

(11) $72 \div 6 =$

(12) $45 \div 9 =$

(13) $20 \div 4 =$

(14) $16 \div 8 =$

(15) $36 \div 6 =$

(16) $7 \div 1 =$

(17) $88 \div 11 =$

(18) $14 \div 2 =$

(19) $2 \div 2 =$

(20) $96 \div 12 =$

(21) $36 \div 4 =$

(22) $4 \div 2 =$

(23) $16 \div 4 =$

(24) $6 \div 2 =$

(25) $144 \div 12 =$

(26) $10 \div 5 =$

(27) $30 \div 5 =$

(28) $11 \div 1 =$

(29) $15 \div 5 =$

(30) $99 \div 9 =$

(31) $24 \div 8 =$

(32) $80 \div 8 =$

(33) $18 \div 2 =$

(34) $24 \div 2 =$

(35) $84 \div 7 =$

(36) $88 \div 8 =$

(37) $132 \div 11 =$

(38) $25 \div 5 =$

(39) $54 \div 6 =$

(40) $96 \div 8 =$

(41) $3 \div 1 =$

(42) $56 \div 7 =$

(43) $12 \div 6 =$

(44) $20 \div 10 =$

(45) $12 \div 12 =$

(46) $1 \div 1 =$

(47) $24 \div 4 =$

(48) $20 \div 5 =$

(49) $12 \div 2 =$

(50) $48 \div 4 =$

(51) $10 \div 5 =$

(52) $72 \div 12 =$

(53) $132 \div 12 =$

(54) $10 \div 5 =$

(55) $20 \div 10 =$

(56) $10 \div 10 =$

(57) $56 \div 7 =$

(58) $81 \div 9 =$

(59) $33 \div 11 =$

(60) $6 \div 1 =$

(1) 44 ÷ 11 =

(2) 30 ÷ 3 =

(3) 16 ÷ 8 =

(4) 36 ÷ 9 =

(5) 36 ÷ 12 =

(6) 54 ÷ 6 =

(7) 8 ÷ 8 =

(8) 48 ÷ 12 =

(9) 56 ÷ 7 =

(10) 30 ÷ 3 =

(11) 100 ÷ 10 =

(12) 54 ÷ 9 =

(13) 18 ÷ 9 =

(14) 18 ÷ 3 =

(15) 3 ÷ 3 =

(16) 88 ÷ 8 =

(17) 5 ÷ 5 =

(18) 12 ÷ 6 =

(19) 24 ÷ 4 =

(20) 35 ÷ 7 =

(21) 36 ÷ 6 =

(22) 11 ÷ 11 =

(23) 1 ÷ 1 =

(24) 55 ÷ 11 =

(25) 90 ÷ 9 =

(26) 30 ÷ 3 =

(27) 2 ÷ 2 =

(28) 42 ÷ 6 =

(29) 48 ÷ 12 =

(30) 55 ÷ 5 =

(31) 45 ÷ 9 =

(32) 48 ÷ 4 =

(33) 9 ÷ 3 =

(34) 9 ÷ 3 =

(35) 6 ÷ 3 =

(36) 8 ÷ 4 =

(37) 56 ÷ 7 =

(38) 32 ÷ 4 =

(39) 30 ÷ 6 =

(40) 5 ÷ 5 =

(41) 60 ÷ 12 =

(42) 3 ÷ 3 =

(43) 16 ÷ 4 =

(44) 80 ÷ 10 =

(45) 10 ÷ 2 =

(46) 12 ÷ 3 =

(47) 96 ÷ 8 =

(48) 16 ÷ 2 =

(49) 63 ÷ 9 =

(50) 42 ÷ 6 =

(51) 72 ÷ 12 =

(52) 12 ÷ 6 =

(53) 45 ÷ 9 =

(54) 2 ÷ 1 =

(55) 33 ÷ 11 =

(56) 30 ÷ 6 =

(57) 63 ÷ 7 =

(58) 63 ÷ 7 =

(59) 48 ÷ 4 =

(60) 16 ÷ 2 =

(1) $40 \div 5 =$

(2) $48 \div 6 =$

(3) $20 \div 4 =$

(4) $120 \div 10 =$

(5) $10 \div 5 =$

(6) $64 \div 8 =$

(7) $4 \div 1 =$

(8) $90 \div 9 =$

(9) $24 \div 4 =$

(10) $7 \div 7 =$

(11) $110 \div 11 =$

(12) $64 \div 8 =$

(13) $30 \div 10 =$

(14) $35 \div 7 =$

(15) $25 \div 5 =$

(16) $12 \div 4 =$

(17) $6 \div 3 =$

(18) $32 \div 8 =$

(19) $8 \div 8 =$

(20) $88 \div 11 =$

(21) $10 \div 5 =$

(22) $1 \div 1 =$

(23) $14 \div 2 =$

(24) $49 \div 7 =$

(25) $14 \div 2 =$

(26) $12 \div 1 =$

(27) $18 \div 2 =$

(28) $77 \div 7 =$

(29) $20 \div 2 =$

(30) $50 \div 10 =$

(31) $144 \div 12 =$

(32) $32 \div 4 =$

(33) $96 \div 8 =$

(34) $55 \div 11 =$

(35) $4 \div 4 =$

(36) $8 \div 4 =$

(37) $108 \div 9 =$

(38) $27 \div 3 =$

(39) $8 \div 2 =$

(40) $77 \div 11 =$

(41) $63 \div 9 =$

(42) $21 \div 3 =$

(43) $99 \div 9 =$

(44) $12 \div 1 =$

(45) $24 \div 6 =$

(46) $56 \div 7 =$

(47) $6 \div 1 =$

(48) $80 \div 8 =$

(49) $80 \div 10 =$

(50) $24 \div 2 =$

(51) $24 \div 12 =$

(52) $100 \div 10 =$

(53) $6 \div 1 =$

(54) $22 \div 2 =$

(55) $121 \div 11 =$

(56) $56 \div 8 =$

(57) $1 \div 1 =$

(58) $55 \div 5 =$

(59) $90 \div 10 =$

(60) $32 \div 8 =$

(1) $30 \div 6 =$

(2) $48 \div 8 =$

(3) $30 \div 6 =$

(4) $40 \div 5 =$

(5) $48 \div 8 =$

(6) $20 \div 5 =$

(7) $99 \div 9 =$

(8) $54 \div 6 =$

(9) $77 \div 7 =$

(10) $7 \div 7 =$

(11) $15 \div 3 =$

(12) $7 \div 7 =$

(13) $12 \div 3 =$

(14) $16 \div 8 =$

(15) $6 \div 1 =$

(16) $18 \div 9 =$

(17) $10 \div 5 =$

(18) $50 \div 10 =$

(19) $24 \div 6 =$

(20) $42 \div 6 =$

(21) $108 \div 12 =$

(22) $36 \div 3 =$

(23) $40 \div 5 =$

(24) $33 \div 11 =$

(25) $9 \div 3 =$

(26) $90 \div 9 =$

(27) $121 \div 11 =$

(28) $12 \div 4 =$

(29) $45 \div 5 =$

(30) $100 \div 10 =$

(31) $8 \div 2 =$

(32) $2 \div 2 =$

(33) $10 \div 10 =$

(34) $18 \div 2 =$

(35) $108 \div 12 =$

(36) $10 \div 2 =$

(37) $55 \div 5 =$

(38) $18 \div 3 =$

(39) $9 \div 1 =$

(40) $6 \div 1 =$

(41) $6 \div 2 =$

(42) $3 \div 1 =$

(43) $63 \div 9 =$

(44) $4 \div 1 =$

(45) $50 \div 10 =$

(46) $132 \div 11 =$

(47) $22 \div 2 =$

(48) $108 \div 12 =$

(49) $84 \div 7 =$

(50) $8 \div 4 =$

(51) $72 \div 6 =$

(52) $28 \div 7 =$

(53) $8 \div 8 =$

(54) $10 \div 5 =$

(55) $132 \div 11 =$

(56) $1 \div 1 =$

(57) $120 \div 12 =$

(58) $108 \div 12 =$

(59) $28 \div 4 =$

(60) $3 \div 1 =$

(1) 44 ÷ 4 =

(2) 33 ÷ 11 =

(3) 120 ÷ 10 =

(4) 80 ÷ 8 =

(5) 36 ÷ 9 =

(6) 8 ÷ 8 =

(7) 27 ÷ 9 =

(8) 121 ÷ 11 =

(9) 44 ÷ 11 =

(10) 9 ÷ 3 =

(11) 54 ÷ 9 =

(12) 35 ÷ 5 =

(13) 120 ÷ 10 =

(14) 12 ÷ 12 =

(15) 5 ÷ 1 =

(16) 18 ÷ 3 =

(17) 63 ÷ 9 =

(18) 5 ÷ 1 =

(19) 12 ÷ 6 =

(20) 30 ÷ 6 =

(21) 88 ÷ 11 =

(22) 81 ÷ 9 =

(23) 36 ÷ 6 =

(24) 10 ÷ 10 =

(25) 100 ÷ 10 =

(26) 84 ÷ 12 =

(27) 12 ÷ 12 =

(28) 25 ÷ 5 =

(29) 12 ÷ 6 =

(30) 14 ÷ 2 =

(31) 60 ÷ 10 =

(32) 27 ÷ 9 =

(33) 12 ÷ 6 =

(34) 49 ÷ 7 =

(35) 20 ÷ 2 =

(36) 44 ÷ 4 =

(37) 84 ÷ 12 =

(38) 88 ÷ 11 =

(39) 30 ÷ 10 =

(40) 6 ÷ 3 =

(41) 90 ÷ 10 =

(42) 40 ÷ 10 =

(43) 108 ÷ 9 =

(44) 77 ÷ 7 =

(45) 108 ÷ 9 =

(46) 30 ÷ 3 =

(47) 120 ÷ 12 =

(48) 60 ÷ 10 =

(49) 16 ÷ 4 =

(50) 144 ÷ 12 =

(51) 24 ÷ 12 =

(52) 11 ÷ 11 =

(53) 8 ÷ 1 =

(54) 30 ÷ 5 =

(55) 8 ÷ 8 =

(56) 99 ÷ 9 =

(57) 40 ÷ 5 =

(58) 14 ÷ 2 =

(59) 110 ÷ 10 =

(60) 144 ÷ 12 =

(1) $9 \div 1 =$

(2) $96 \div 12 =$

(3) $9 \div 9 =$

(4) $108 \div 12 =$

(5) $30 \div 6 =$

(6) $21 \div 3 =$

(7) $84 \div 12 =$

(8) $40 \div 10 =$

(9) $27 \div 3 =$

(10) $60 \div 6 =$

(11) $28 \div 7 =$

(12) $22 \div 2 =$

(13) $10 \div 2 =$

(14) $20 \div 5 =$

(15) $72 \div 9 =$

(16) $108 \div 9 =$

(17) $28 \div 7 =$

(18) $54 \div 9 =$

(19) $54 \div 9 =$

(20) $44 \div 4 =$

(21) $66 \div 6 =$

(22) $66 \div 11 =$

(23) $10 \div 5 =$

(24) $6 \div 2 =$

(25) $20 \div 4 =$

(26) $11 \div 11 =$

(27) $36 \div 3 =$

(28) $36 \div 6 =$

(29) $27 \div 3 =$

(30) $20 \div 10 =$

(31) $5 \div 1 =$

(32) $24 \div 3 =$

(33) $45 \div 5 =$

(34) $10 \div 10 =$

(35) $55 \div 11 =$

(36) $110 \div 11 =$

(37) $24 \div 6 =$

(38) $6 \div 6 =$

(39) $45 \div 9 =$

(40) $32 \div 8 =$

(41) $24 \div 12 =$

(42) $11 \div 1 =$

(43) $66 \div 11 =$

(44) $42 \div 6 =$

(45) $42 \div 7 =$

(46) $40 \div 5 =$

(47) $81 \div 9 =$

(48) $56 \div 8 =$

(49) $36 \div 6 =$

(50) $4 \div 2 =$

(51) $49 \div 7 =$

(52) $99 \div 9 =$

(53) $72 \div 9 =$

(54) $99 \div 9 =$

(55) $4 \div 4 =$

(56) $64 \div 8 =$

(57) $36 \div 9 =$

(58) $16 \div 8 =$

(59) $84 \div 12 =$

(60) $40 \div 4 =$

(1) $55 \div 11 =$

(2) $11 \div 11 =$

(3) $36 \div 6 =$

(4) $50 \div 5 =$

(5) $24 \div 4 =$

(6) $72 \div 9 =$

(7) $28 \div 4 =$

(8) $88 \div 8 =$

(9) $3 \div 1 =$

(10) $18 \div 6 =$

(11) $24 \div 8 =$

(12) $54 \div 6 =$

(13) $14 \div 2 =$

(14) $121 \div 11 =$

(15) $5 \div 5 =$

(16) $21 \div 3 =$

(17) $8 \div 8 =$

(18) $60 \div 5 =$

(19) $33 \div 3 =$

(20) $4 \div 4 =$

(21) $40 \div 8 =$

(22) $32 \div 8 =$

(23) $4 \div 2 =$

(24) $22 \div 11 =$

(25) $27 \div 9 =$

(26) $18 \div 9 =$

(27) $60 \div 6 =$

(28) $54 \div 6 =$

(29) $54 \div 9 =$

(30) $100 \div 10 =$

(31) $5 \div 1 =$

(32) $32 \div 4 =$

(33) $9 \div 3 =$

(34) $20 \div 4 =$

(35) $21 \div 7 =$

(36) $108 \div 12 =$

(37) $32 \div 8 =$

(38) $6 \div 6 =$

(39) $110 \div 11 =$

(40) $30 \div 6 =$

(41) $42 \div 6 =$

(42) $7 \div 7 =$

(43) $35 \div 7 =$

(44) $21 \div 7 =$

(45) $90 \div 9 =$

(46) $132 \div 11 =$

(47) $18 \div 9 =$

(48) $2 \div 1 =$

(49) $120 \div 10 =$

(50) $70 \div 10 =$

(51) $60 \div 6 =$

(52) $12 \div 1 =$

(53) $108 \div 9 =$

(54) $99 \div 9 =$

(55) $6 \div 2 =$

(56) $88 \div 8 =$

(57) $72 \div 9 =$

(58) $50 \div 10 =$

(59) $6 \div 6 =$

(60) $20 \div 4 =$

(1) $45 \div 9 =$

(2) $18 \div 3 =$

(3) $36 \div 12 =$

(4) $110 \div 11 =$

(5) $21 \div 3 =$

(6) $63 \div 7 =$

(7) $24 \div 2 =$

(8) $72 \div 8 =$

(9) $24 \div 2 =$

(10) $14 \div 7 =$

(11) $80 \div 8 =$

(12) $20 \div 5 =$

(13) $11 \div 1 =$

(14) $11 \div 11 =$

(15) $144 \div 12 =$

(16) $144 \div 12 =$

(17) $9 \div 3 =$

(18) $120 \div 12 =$

(19) $18 \div 3 =$

(20) $108 \div 12 =$

(21) $8 \div 4 =$

(22) $24 \div 2 =$

(23) $16 \div 8 =$

(24) $60 \div 5 =$

(25) $12 \div 1 =$

(26) $6 \div 1 =$

(27) $16 \div 8 =$

(28) $81 \div 9 =$

(29) $5 \div 5 =$

(30) $54 \div 6 =$

(31) $63 \div 7 =$

(32) $5 \div 5 =$

(33) $66 \div 6 =$

(34) $120 \div 12 =$

(35) $80 \div 10 =$

(36) $14 \div 7 =$

(37) $3 \div 1 =$

(38) $110 \div 11 =$

(39) $36 \div 9 =$

(40) $40 \div 4 =$

(41) $7 \div 7 =$

(42) $2 \div 1 =$

(43) $10 \div 2 =$

(44) $4 \div 2 =$

(45) $30 \div 10 =$

(46) $132 \div 11 =$

(47) $35 \div 7 =$

(48) $40 \div 5 =$

(49) $16 \div 8 =$

(50) $32 \div 4 =$

(51) $48 \div 12 =$

(52) $10 \div 10 =$

(53) $15 \div 5 =$

(54) $40 \div 4 =$

(55) $66 \div 11 =$

(56) $3 \div 3 =$

(57) $72 \div 6 =$

(58) $81 \div 9 =$

(59) $60 \div 5 =$

(60) $8 \div 4 =$

(1) 27 ÷ 3 =   (2) 60 ÷ 10 =   (3) 36 ÷ 3 =

(4) 24 ÷ 12 =   (5) 11 ÷ 1 =   (6) 56 ÷ 8 =

(7) 48 ÷ 6 =   (8) 54 ÷ 6 =   (9) 18 ÷ 2 =

(10) 16 ÷ 2 =   (11) 99 ÷ 11 =   (12) 48 ÷ 12 =

(13) 36 ÷ 9 =   (14) 36 ÷ 3 =   (15) 20 ÷ 4 =

(16) 20 ÷ 5 =   (17) 44 ÷ 4 =   (18) 40 ÷ 4 =

(19) 66 ÷ 11 =   (20) 27 ÷ 9 =   (21) 16 ÷ 8 =

(22) 36 ÷ 4 =   (23) 81 ÷ 9 =   (24) 1 ÷ 1 =

(25) 6 ÷ 6 =   (26) 99 ÷ 11 =   (27) 24 ÷ 3 =

(28) 9 ÷ 1 =   (29) 49 ÷ 7 =   (30) 77 ÷ 7 =

(31) 2 ÷ 2 =   (32) 28 ÷ 7 =   (33) 24 ÷ 6 =

(34) 20 ÷ 2 =   (35) 3 ÷ 1 =   (36) 60 ÷ 10 =

(37) 8 ÷ 8 =   (38) 56 ÷ 7 =   (39) 45 ÷ 5 =

(40) 81 ÷ 9 =   (41) 36 ÷ 9 =   (42) 21 ÷ 3 =

(43) 32 ÷ 4 =   (44) 22 ÷ 2 =   (45) 121 ÷ 11 =

(46) 18 ÷ 6 =   (47) 33 ÷ 3 =   (48) 60 ÷ 12 =

(49) 132 ÷ 12 =   (50) 108 ÷ 9 =   (51) 72 ÷ 12 =

(52) 11 ÷ 1 =   (53) 9 ÷ 9 =   (54) 18 ÷ 6 =

(55) 15 ÷ 5 =   (56) 110 ÷ 10 =   (57) 11 ÷ 1 =

(58) 40 ÷ 5 =   (59) 99 ÷ 9 =   (60) 10 ÷ 10 =

(1) $60 \div 5 =$

(2) $21 \div 3 =$

(3) $36 \div 9 =$

(4) $45 \div 5 =$

(5) $24 \div 6 =$

(6) $77 \div 7 =$

(7) $20 \div 2 =$

(8) $36 \div 9 =$

(9) $6 \div 2 =$

(10) $15 \div 3 =$

(11) $81 \div 9 =$

(12) $12 \div 12 =$

(13) $30 \div 10 =$

(14) $54 \div 9 =$

(15) $90 \div 9 =$

(16) $20 \div 5 =$

(17) $10 \div 5 =$

(18) $16 \div 4 =$

(19) $14 \div 7 =$

(20) $48 \div 8 =$

(21) $70 \div 10 =$

(22) $120 \div 12 =$

(23) $2 \div 1 =$

(24) $21 \div 3 =$

(25) $10 \div 10 =$

(26) $8 \div 8 =$

(27) $63 \div 7 =$

(28) $30 \div 10 =$

(29) $32 \div 8 =$

(30) $108 \div 12 =$

(31) $45 \div 9 =$

(32) $60 \div 10 =$

(33) $132 \div 12 =$

(34) $54 \div 9 =$

(35) $10 \div 1 =$

(36) $80 \div 8 =$

(37) $50 \div 10 =$

(38) $72 \div 12 =$

(39) $55 \div 5 =$

(40) $36 \div 12 =$

(41) $84 \div 7 =$

(42) $120 \div 10 =$

(43) $16 \div 8 =$

(44) $20 \div 10 =$

(45) $30 \div 5 =$

(46) $60 \div 6 =$

(47) $36 \div 12 =$

(48) $16 \div 4 =$

(49) $54 \div 6 =$

(50) $8 \div 1 =$

(51) $16 \div 2 =$

(52) $110 \div 10 =$

(53) $120 \div 10 =$

(54) $99 \div 11 =$

(55) $3 \div 1 =$

(56) $60 \div 12 =$

(57) $55 \div 5 =$

(58) $1 \div 1 =$

(59) $8 \div 1 =$

(60) $7 \div 1 =$

(1) 35 ÷ 5 =

(2) 12 ÷ 3 =

(3) 60 ÷ 6 =

(4) 30 ÷ 3 =

(5) 8 ÷ 2 =

(6) 27 ÷ 3 =

(7) 66 ÷ 11 =

(8) 80 ÷ 8 =

(9) 30 ÷ 6 =

(10) 15 ÷ 5 =

(11) 144 ÷ 12 =

(12) 7 ÷ 1 =

(13) 12 ÷ 2 =

(14) 121 ÷ 11 =

(15) 84 ÷ 7 =

(16) 35 ÷ 7 =

(17) 45 ÷ 9 =

(18) 40 ÷ 8 =

(19) 15 ÷ 3 =

(20) 50 ÷ 10 =

(21) 110 ÷ 10 =

(22) 77 ÷ 11 =

(23) 72 ÷ 8 =

(24) 24 ÷ 8 =

(25) 7 ÷ 7 =

(26) 84 ÷ 7 =

(27) 8 ÷ 2 =

(28) 11 ÷ 11 =

(29) 6 ÷ 3 =

(30) 80 ÷ 10 =

(31) 14 ÷ 7 =

(32) 40 ÷ 8 =

(33) 6 ÷ 6 =

(34) 28 ÷ 7 =

(35) 132 ÷ 12 =

(36) 99 ÷ 9 =

(37) 8 ÷ 8 =

(38) 56 ÷ 7 =

(39) 121 ÷ 11 =

(40) 30 ÷ 10 =

(41) 60 ÷ 6 =

(42) 25 ÷ 5 =

(43) 10 ÷ 1 =

(44) 16 ÷ 2 =

(45) 6 ÷ 3 =

(46) 6 ÷ 6 =

(47) 33 ÷ 11 =

(48) 14 ÷ 2 =

(49) 14 ÷ 7 =

(50) 15 ÷ 5 =

(51) 64 ÷ 8 =

(52) 32 ÷ 8 =

(53) 40 ÷ 8 =

(54) 7 ÷ 1 =

(55) 36 ÷ 12 =

(56) 48 ÷ 6 =

(57) 33 ÷ 3 =

(58) 10 ÷ 1 =

(59) 42 ÷ 7 =

(60) 33 ÷ 3 =

| Day: | 67 | Date: | | Score: | /60 |
| Name: | | Time: | : | Rating: | ☆☆☆☆☆ |

(1) 24 ÷ 4 =

(2) 96 ÷ 12 =

(3) 28 ÷ 4 =

(4) 15 ÷ 3 =

(5) 66 ÷ 11 =

(6) 40 ÷ 5 =

(7) 22 ÷ 11 =

(8) 84 ÷ 12 =

(9) 72 ÷ 9 =

(10) 36 ÷ 9 =

(11) 55 ÷ 11 =

(12) 36 ÷ 9 =

(13) 4 ÷ 1 =

(14) 90 ÷ 9 =

(15) 10 ÷ 1 =

(16) 36 ÷ 9 =

(17) 10 ÷ 1 =

(18) 24 ÷ 3 =

(19) 18 ÷ 2 =

(20) 88 ÷ 11 =

(21) 121 ÷ 11 =

(22) 84 ÷ 12 =

(23) 80 ÷ 10 =

(24) 72 ÷ 6 =

(25) 48 ÷ 12 =

(26) 9 ÷ 3 =

(27) 25 ÷ 5 =

(28) 22 ÷ 11 =

(29) 108 ÷ 9 =

(30) 18 ÷ 3 =

(31) 77 ÷ 7 =

(32) 12 ÷ 2 =

(33) 66 ÷ 11 =

(34) 14 ÷ 7 =

(35) 12 ÷ 1 =

(36) 8 ÷ 1 =

(37) 70 ÷ 7 =

(38) 24 ÷ 4 =

(39) 9 ÷ 3 =

(40) 50 ÷ 5 =

(41) 4 ÷ 4 =

(42) 120 ÷ 10 =

(43) 48 ÷ 4 =

(44) 6 ÷ 6 =

(45) 60 ÷ 10 =

(46) 88 ÷ 8 =

(47) 70 ÷ 10 =

(48) 12 ÷ 2 =

(49) 8 ÷ 4 =

(50) 6 ÷ 3 =

(51) 44 ÷ 4 =

(52) 90 ÷ 10 =

(53) 60 ÷ 12 =

(54) 25 ÷ 5 =

(55) 6 ÷ 1 =

(56) 12 ÷ 3 =

(57) 5 ÷ 1 =

(58) 21 ÷ 7 =

(59) 55 ÷ 5 =

(60) 36 ÷ 12 =

(1) $10 \div 1 =$　　(2) $4 \div 2 =$　　(3) $8 \div 4 =$

(4) $24 \div 3 =$　　(5) $8 \div 1 =$　　(6) $6 \div 2 =$

(7) $72 \div 6 =$　　(8) $42 \div 7 =$　　(9) $40 \div 5 =$

(10) $48 \div 6 =$　　(11) $10 \div 2 =$　　(12) $121 \div 11 =$

(13) $35 \div 7 =$　　(14) $40 \div 4 =$　　(15) $36 \div 6 =$

(16) $90 \div 10 =$　　(17) $36 \div 6 =$　　(18) $12 \div 12 =$

(19) $22 \div 11 =$　　(20) $33 \div 3 =$　　(21) $20 \div 10 =$

(22) $12 \div 12 =$　　(23) $15 \div 5 =$　　(24) $48 \div 12 =$

(25) $42 \div 6 =$　　(26) $84 \div 7 =$　　(27) $54 \div 9 =$

(28) $12 \div 3 =$　　(29) $11 \div 1 =$　　(30) $24 \div 12 =$

(31) $24 \div 6 =$　　(32) $55 \div 5 =$　　(33) $54 \div 9 =$

(34) $20 \div 5 =$　　(35) $72 \div 6 =$　　(36) $30 \div 3 =$

(37) $2 \div 1 =$　　(38) $11 \div 1 =$　　(39) $90 \div 9 =$

(40) $6 \div 1 =$　　(41) $36 \div 3 =$　　(42) $110 \div 10 =$

(43) $27 \div 3 =$　　(44) $77 \div 7 =$　　(45) $24 \div 3 =$

(46) $24 \div 12 =$　　(47) $132 \div 12 =$　　(48) $99 \div 11 =$

(49) $18 \div 3 =$　　(50) $60 \div 5 =$　　(51) $48 \div 4 =$

(52) $27 \div 3 =$　　(53) $72 \div 9 =$　　(54) $30 \div 6 =$

(55) $132 \div 11 =$　　(56) $8 \div 2 =$　　(57) $44 \div 11 =$

(58) $36 \div 4 =$　　(59) $20 \div 10 =$　　(60) $24 \div 8 =$

(1) $66 \div 11 =$

(2) $48 \div 8 =$

(3) $108 \div 9 =$

(4) $56 \div 7 =$

(5) $30 \div 6 =$

(6) $10 \div 10 =$

(7) $22 \div 11 =$

(8) $27 \div 9 =$

(9) $49 \div 7 =$

(10) $56 \div 8 =$

(11) $24 \div 6 =$

(12) $56 \div 8 =$

(13) $48 \div 4 =$

(14) $66 \div 6 =$

(15) $28 \div 4 =$

(16) $132 \div 12 =$

(17) $30 \div 6 =$

(18) $14 \div 2 =$

(19) $45 \div 5 =$

(20) $4 \div 2 =$

(21) $110 \div 11 =$

(22) $15 \div 5 =$

(23) $4 \div 2 =$

(24) $15 \div 3 =$

(25) $56 \div 8 =$

(26) $21 \div 7 =$

(27) $4 \div 1 =$

(28) $54 \div 9 =$

(29) $72 \div 8 =$

(30) $10 \div 5 =$

(31) $11 \div 1 =$

(32) $88 \div 8 =$

(33) $108 \div 9 =$

(34) $44 \div 4 =$

(35) $36 \div 4 =$

(36) $4 \div 4 =$

(37) $28 \div 4 =$

(38) $10 \div 1 =$

(39) $45 \div 9 =$

(40) $4 \div 2 =$

(41) $20 \div 4 =$

(42) $11 \div 1 =$

(43) $88 \div 8 =$

(44) $2 \div 1 =$

(45) $99 \div 9 =$

(46) $24 \div 3 =$

(47) $80 \div 8 =$

(48) $16 \div 2 =$

(49) $24 \div 6 =$

(50) $11 \div 11 =$

(51) $90 \div 10 =$

(52) $21 \div 7 =$

(53) $24 \div 4 =$

(54) $63 \div 7 =$

(55) $33 \div 3 =$

(56) $18 \div 6 =$

(57) $12 \div 4 =$

(58) $77 \div 11 =$

(59) $15 \div 3 =$

(60) $40 \div 5 =$

(1) $99 \div 9 =$

(2) $63 \div 9 =$

(3) $10 \div 2 =$

(4) $36 \div 6 =$

(5) $21 \div 7 =$

(6) $22 \div 2 =$

(7) $96 \div 12 =$

(8) $84 \div 12 =$

(9) $35 \div 5 =$

(10) $55 \div 5 =$

(11) $16 \div 2 =$

(12) $64 \div 8 =$

(13) $60 \div 10 =$

(14) $60 \div 12 =$

(15) $5 \div 1 =$

(16) $72 \div 8 =$

(17) $20 \div 10 =$

(18) $72 \div 6 =$

(19) $63 \div 7 =$

(20) $12 \div 2 =$

(21) $50 \div 5 =$

(22) $70 \div 10 =$

(23) $6 \div 2 =$

(24) $50 \div 5 =$

(25) $8 \div 1 =$

(26) $55 \div 5 =$

(27) $40 \div 10 =$

(28) $55 \div 11 =$

(29) $9 \div 3 =$

(30) $72 \div 8 =$

(31) $18 \div 2 =$

(32) $12 \div 12 =$

(33) $7 \div 1 =$

(34) $32 \div 4 =$

(35) $55 \div 11 =$

(36) $70 \div 7 =$

(37) $15 \div 5 =$

(38) $49 \div 7 =$

(39) $84 \div 7 =$

(40) $40 \div 10 =$

(41) $24 \div 6 =$

(42) $45 \div 5 =$

(43) $40 \div 10 =$

(44) $28 \div 7 =$

(45) $120 \div 12 =$

(46) $7 \div 7 =$

(47) $72 \div 12 =$

(48) $16 \div 4 =$

(49) $120 \div 12 =$

(50) $72 \div 12 =$

(51) $36 \div 3 =$

(52) $2 \div 2 =$

(53) $12 \div 6 =$

(54) $20 \div 4 =$

(55) $45 \div 5 =$

(56) $30 \div 10 =$

(57) $32 \div 4 =$

(58) $14 \div 2 =$

(59) $36 \div 12 =$

(60) $15 \div 3 =$

(1) $3 \div 3 =$

(2) $120 \div 12 =$

(3) $21 \div 7 =$

(4) $33 \div 11 =$

(5) $110 \div 10 =$

(6) $120 \div 10 =$

(7) $36 \div 3 =$

(8) $10 \div 5 =$

(9) $24 \div 3 =$

(10) $14 \div 2 =$

(11) $8 \div 8 =$

(12) $90 \div 10 =$

(13) $24 \div 8 =$

(14) $84 \div 12 =$

(15) $22 \div 2 =$

(16) $28 \div 7 =$

(17) $48 \div 8 =$

(18) $28 \div 4 =$

(19) $2 \div 2 =$

(20) $6 \div 2 =$

(21) $12 \div 3 =$

(22) $10 \div 10 =$

(23) $21 \div 7 =$

(24) $6 \div 6 =$

(25) $11 \div 11 =$

(26) $40 \div 10 =$

(27) $49 \div 7 =$

(28) $33 \div 3 =$

(29) $72 \div 12 =$

(30) $24 \div 8 =$

(31) $48 \div 8 =$

(32) $4 \div 4 =$

(33) $24 \div 2 =$

(34) $35 \div 5 =$

(35) $132 \div 11 =$

(36) $18 \div 6 =$

(37) $44 \div 11 =$

(38) $45 \div 9 =$

(39) $64 \div 8 =$

(40) $28 \div 7 =$

(41) $80 \div 10 =$

(42) $20 \div 4 =$

(43) $99 \div 11 =$

(44) $90 \div 9 =$

(45) $6 \div 3 =$

(46) $6 \div 2 =$

(47) $8 \div 4 =$

(48) $110 \div 10 =$

(49) $15 \div 5 =$

(50) $6 \div 3 =$

(51) $4 \div 4 =$

(52) $40 \div 8 =$

(53) $33 \div 3 =$

(54) $48 \div 4 =$

(55) $22 \div 2 =$

(56) $48 \div 12 =$

(57) $4 \div 2 =$

(58) $30 \div 10 =$

(59) $7 \div 1 =$

(60) $6 \div 1 =$

(1) $30 \div 10 =$

(2) $132 \div 12 =$

(3) $12 \div 6 =$

(4) $20 \div 5 =$

(5) $27 \div 3 =$

(6) $24 \div 2 =$

(7) $9 \div 3 =$

(8) $80 \div 8 =$

(9) $96 \div 12 =$

(10) $64 \div 8 =$

(11) $9 \div 9 =$

(12) $33 \div 3 =$

(13) $100 \div 10 =$

(14) $48 \div 6 =$

(15) $12 \div 4 =$

(16) $55 \div 5 =$

(17) $50 \div 5 =$

(18) $108 \div 12 =$

(19) $20 \div 10 =$

(20) $30 \div 6 =$

(21) $35 \div 7 =$

(22) $24 \div 2 =$

(23) $22 \div 2 =$

(24) $30 \div 5 =$

(25) $70 \div 7 =$

(26) $11 \div 1 =$

(27) $7 \div 7 =$

(28) $108 \div 12 =$

(29) $18 \div 3 =$

(30) $3 \div 3 =$

(31) $18 \div 2 =$

(32) $88 \div 8 =$

(33) $2 \div 2 =$

(34) $16 \div 2 =$

(35) $108 \div 9 =$

(36) $40 \div 8 =$

(37) $42 \div 7 =$

(38) $16 \div 4 =$

(39) $45 \div 5 =$

(40) $11 \div 1 =$

(41) $2 \div 2 =$

(42) $1 \div 1 =$

(43) $100 \div 10 =$

(44) $90 \div 10 =$

(45) $120 \div 12 =$

(46) $32 \div 4 =$

(47) $10 \div 1 =$

(48) $48 \div 8 =$

(49) $60 \div 5 =$

(50) $9 \div 1 =$

(51) $64 \div 8 =$

(52) $27 \div 9 =$

(53) $40 \div 8 =$

(54) $66 \div 6 =$

(55) $132 \div 12 =$

(56) $50 \div 5 =$

(57) $77 \div 11 =$

(58) $144 \div 12 =$

(59) $2 \div 1 =$

(60) $48 \div 4 =$

(1) 44 ÷ 11 =

(2) 33 ÷ 11 =

(3) 10 ÷ 1 =

(4) 121 ÷ 11 =

(5) 64 ÷ 8 =

(6) 18 ÷ 9 =

(7) 3 ÷ 3 =

(8) 88 ÷ 11 =

(9) 20 ÷ 2 =

(10) 22 ÷ 11 =

(11) 72 ÷ 12 =

(12) 24 ÷ 8 =

(13) 88 ÷ 8 =

(14) 144 ÷ 12 =

(15) 40 ÷ 5 =

(16) 7 ÷ 1 =

(17) 40 ÷ 5 =

(18) 12 ÷ 6 =

(19) 60 ÷ 5 =

(20) 36 ÷ 12 =

(21) 56 ÷ 7 =

(22) 12 ÷ 3 =

(23) 30 ÷ 3 =

(24) 10 ÷ 5 =

(25) 9 ÷ 9 =

(26) 42 ÷ 7 =

(27) 20 ÷ 5 =

(28) 44 ÷ 4 =

(29) 54 ÷ 9 =

(30) 2 ÷ 1 =

(31) 88 ÷ 11 =

(32) 15 ÷ 3 =

(33) 110 ÷ 10 =

(34) 12 ÷ 6 =

(35) 132 ÷ 12 =

(36) 63 ÷ 9 =

(37) 6 ÷ 2 =

(38) 30 ÷ 5 =

(39) 40 ÷ 4 =

(40) 8 ÷ 2 =

(41) 63 ÷ 9 =

(42) 25 ÷ 5 =

(43) 24 ÷ 4 =

(44) 48 ÷ 12 =

(45) 36 ÷ 4 =

(46) 60 ÷ 12 =

(47) 10 ÷ 2 =

(48) 24 ÷ 8 =

(49) 99 ÷ 11 =

(50) 56 ÷ 7 =

(51) 84 ÷ 12 =

(52) 27 ÷ 3 =

(53) 25 ÷ 5 =

(54) 20 ÷ 5 =

(55) 30 ÷ 6 =

(56) 48 ÷ 4 =

(57) 7 ÷ 7 =

(58) 72 ÷ 6 =

(59) 21 ÷ 7 =

(60) 12 ÷ 12 =

Day:	74	Date:		Score:	/60
Name:		Time:	:	Rating:	☆☆☆☆☆

(1) $10 \div 10 =$

(2) $28 \div 4 =$

(3) $40 \div 10 =$

(4) $5 \div 5 =$

(5) $90 \div 10 =$

(6) $8 \div 1 =$

(7) $36 \div 9 =$

(8) $10 \div 2 =$

(9) $66 \div 6 =$

(10) $12 \div 4 =$

(11) $90 \div 10 =$

(12) $28 \div 7 =$

(13) $110 \div 10 =$

(14) $81 \div 9 =$

(15) $45 \div 5 =$

(16) $96 \div 8 =$

(17) $49 \div 7 =$

(18) $120 \div 12 =$

(19) $88 \div 11 =$

(20) $42 \div 7 =$

(21) $36 \div 3 =$

(22) $21 \div 7 =$

(23) $36 \div 6 =$

(24) $18 \div 2 =$

(25) $10 \div 1 =$

(26) $72 \div 8 =$

(27) $20 \div 10 =$

(28) $12 \div 2 =$

(29) $3 \div 1 =$

(30) $54 \div 6 =$

(31) $72 \div 9 =$

(32) $8 \div 4 =$

(33) $16 \div 8 =$

(34) $25 \div 5 =$

(35) $10 \div 5 =$

(36) $108 \div 9 =$

(37) $9 \div 3 =$

(38) $45 \div 9 =$

(39) $63 \div 9 =$

(40) $10 \div 10 =$

(41) $27 \div 9 =$

(42) $11 \div 11 =$

(43) $72 \div 6 =$

(44) $20 \div 2 =$

(45) $1 \div 1 =$

(46) $64 \div 8 =$

(47) $6 \div 2 =$

(48) $77 \div 11 =$

(49) $36 \div 4 =$

(50) $36 \div 6 =$

(51) $42 \div 7 =$

(52) $36 \div 12 =$

(53) $22 \div 2 =$

(54) $49 \div 7 =$

(55) $4 \div 4 =$

(56) $60 \div 10 =$

(57) $72 \div 12 =$

(58) $18 \div 9 =$

(59) $40 \div 5 =$

(60) $5 \div 5 =$

(1) $12 \div 1 =$

(2) $24 \div 4 =$

(3) $20 \div 2 =$

(4) $63 \div 7 =$

(5) $8 \div 1 =$

(6) $96 \div 12 =$

(7) $11 \div 1 =$

(8) $8 \div 2 =$

(9) $25 \div 5 =$

(10) $6 \div 1 =$

(11) $42 \div 6 =$

(12) $48 \div 8 =$

(13) $5 \div 1 =$

(14) $9 \div 1 =$

(15) $72 \div 12 =$

(16) $15 \div 5 =$

(17) $12 \div 4 =$

(18) $4 \div 1 =$

(19) $30 \div 5 =$

(20) $8 \div 2 =$

(21) $110 \div 10 =$

(22) $60 \div 6 =$

(23) $44 \div 11 =$

(24) $48 \div 12 =$

(25) $108 \div 9 =$

(26) $96 \div 8 =$

(27) $6 \div 6 =$

(28) $1 \div 1 =$

(29) $72 \div 12 =$

(30) $18 \div 6 =$

(31) $28 \div 7 =$

(32) $84 \div 7 =$

(33) $120 \div 10 =$

(34) $96 \div 8 =$

(35) $4 \div 1 =$

(36) $99 \div 11 =$

(37) $12 \div 6 =$

(38) $70 \div 7 =$

(39) $110 \div 11 =$

(40) $36 \div 9 =$

(41) $42 \div 7 =$

(42) $7 \div 7 =$

(43) $32 \div 8 =$

(44) $40 \div 10 =$

(45) $108 \div 12 =$

(46) $12 \div 4 =$

(47) $18 \div 6 =$

(48) $44 \div 4 =$

(49) $8 \div 4 =$

(50) $21 \div 7 =$

(51) $12 \div 3 =$

(52) $55 \div 5 =$

(53) $7 \div 7 =$

(54) $30 \div 6 =$

(55) $120 \div 12 =$

(56) $30 \div 5 =$

(57) $22 \div 2 =$

(58) $44 \div 4 =$

(59) $108 \div 12 =$

(60) $36 \div 6 =$

(1) $27 \div 9 =$

(2) $44 \div 4 =$

(3) $11 \div 11 =$

(4) $50 \div 5 =$

(5) $20 \div 4 =$

(6) $18 \div 3 =$

(7) $90 \div 9 =$

(8) $72 \div 8 =$

(9) $33 \div 3 =$

(10) $16 \div 2 =$

(11) $45 \div 5 =$

(12) $16 \div 2 =$

(13) $25 \div 5 =$

(14) $110 \div 11 =$

(15) $24 \div 3 =$

(16) $6 \div 3 =$

(17) $28 \div 7 =$

(18) $24 \div 2 =$

(19) $30 \div 10 =$

(20) $88 \div 11 =$

(21) $27 \div 3 =$

(22) $4 \div 1 =$

(23) $18 \div 2 =$

(24) $42 \div 7 =$

(25) $9 \div 9 =$

(26) $96 \div 8 =$

(27) $90 \div 9 =$

(28) $27 \div 9 =$

(29) $42 \div 7 =$

(30) $50 \div 5 =$

(31) $30 \div 5 =$

(32) $12 \div 2 =$

(33) $20 \div 2 =$

(34) $12 \div 3 =$

(35) $55 \div 5 =$

(36) $48 \div 12 =$

(37) $55 \div 11 =$

(38) $63 \div 7 =$

(39) $5 \div 1 =$

(40) $40 \div 10 =$

(41) $16 \div 2 =$

(42) $90 \div 9 =$

(43) $24 \div 3 =$

(44) $60 \div 10 =$

(45) $24 \div 2 =$

(46) $121 \div 11 =$

(47) $16 \div 4 =$

(48) $14 \div 2 =$

(49) $5 \div 5 =$

(50) $56 \div 7 =$

(51) $40 \div 5 =$

(52) $54 \div 9 =$

(53) $9 \div 3 =$

(54) $42 \div 6 =$

(55) $132 \div 12 =$

(56) $60 \div 10 =$

(57) $40 \div 4 =$

(58) $63 \div 9 =$

(59) $72 \div 6 =$

(60) $84 \div 7 =$

(1) $12 \div 3 =$ (2) $33 \div 3 =$ (3) $6 \div 6 =$

(4) $77 \div 11 =$ (5) $50 \div 5 =$ (6) $120 \div 12 =$

(7) $96 \div 8 =$ (8) $100 \div 10 =$ (9) $32 \div 4 =$

(10) $60 \div 12 =$ (11) $6 \div 6 =$ (12) $36 \div 12 =$

(13) $30 \div 10 =$ (14) $14 \div 2 =$ (15) $24 \div 6 =$

(16) $27 \div 9 =$ (17) $18 \div 2 =$ (18) $48 \div 6 =$

(19) $36 \div 3 =$ (20) $72 \div 12 =$ (21) $4 \div 1 =$

(22) $30 \div 5 =$ (23) $108 \div 12 =$ (24) $72 \div 9 =$

(25) $10 \div 1 =$ (26) $10 \div 5 =$ (27) $110 \div 10 =$

(28) $108 \div 9 =$ (29) $55 \div 5 =$ (30) $88 \div 11 =$

(31) $18 \div 6 =$ (32) $36 \div 9 =$ (33) $55 \div 5 =$

(34) $77 \div 7 =$ (35) $66 \div 6 =$ (36) $45 \div 5 =$

(37) $32 \div 8 =$ (38) $12 \div 3 =$ (39) $22 \div 11 =$

(40) $21 \div 3 =$ (41) $22 \div 2 =$ (42) $80 \div 8 =$

(43) $99 \div 11 =$ (44) $77 \div 11 =$ (45) $132 \div 12 =$

(46) $1 \div 1 =$ (47) $50 \div 5 =$ (48) $81 \div 9 =$

(49) $24 \div 2 =$ (50) $36 \div 4 =$ (51) $20 \div 4 =$

(52) $7 \div 7 =$ (53) $9 \div 3 =$ (54) $72 \div 12 =$

(55) $28 \div 4 =$ (56) $6 \div 1 =$ (57) $35 \div 5 =$

(58) $8 \div 4 =$ (59) $18 \div 9 =$ (60) $20 \div 4 =$

(1) $35 \div 7 =$ (2) $24 \div 8 =$ (3) $40 \div 10 =$

(4) $80 \div 10 =$ (5) $4 \div 4 =$ (6) $48 \div 8 =$

(7) $6 \div 3 =$ (8) $42 \div 7 =$ (9) $32 \div 8 =$

(10) $8 \div 1 =$ (11) $35 \div 5 =$ (12) $60 \div 6 =$

(13) $49 \div 7 =$ (14) $66 \div 6 =$ (15) $72 \div 9 =$

(16) $132 \div 11 =$ (17) $22 \div 2 =$ (18) $16 \div 2 =$

(19) $27 \div 9 =$ (20) $108 \div 12 =$ (21) $14 \div 2 =$

(22) $27 \div 3 =$ (23) $48 \div 6 =$ (24) $66 \div 6 =$

(25) $36 \div 3 =$ (26) $10 \div 5 =$ (27) $21 \div 7 =$

(28) $77 \div 7 =$ (29) $18 \div 2 =$ (30) $21 \div 7 =$

(31) $16 \div 2 =$ (32) $4 \div 4 =$ (33) $70 \div 10 =$

(34) $48 \div 4 =$ (35) $28 \div 4 =$ (36) $40 \div 10 =$

(37) $9 \div 1 =$ (38) $88 \div 8 =$ (39) $8 \div 4 =$

(40) $99 \div 11 =$ (41) $6 \div 3 =$ (42) $88 \div 8 =$

(43) $77 \div 7 =$ (44) $48 \div 4 =$ (45) $9 \div 9 =$

(46) $18 \div 6 =$ (47) $35 \div 7 =$ (48) $70 \div 10 =$

(49) $40 \div 8 =$ (50) $24 \div 4 =$ (51) $12 \div 2 =$

(52) $45 \div 9 =$ (53) $72 \div 8 =$ (54) $7 \div 1 =$

(55) $20 \div 2 =$ (56) $77 \div 7 =$ (57) $40 \div 5 =$

(58) $48 \div 6 =$ (59) $100 \div 10 =$ (60) $12 \div 3 =$

(1) $120 \div 10 =$ (2) $21 \div 3 =$ (3) $32 \div 8 =$

(4) $30 \div 5 =$ (5) $100 \div 10 =$ (6) $66 \div 6 =$

(7) $72 \div 8 =$ (8) $4 \div 4 =$ (9) $35 \div 5 =$

(10) $54 \div 6 =$ (11) $77 \div 7 =$ (12) $132 \div 12 =$

(13) $11 \div 11 =$ (14) $99 \div 11 =$ (15) $28 \div 7 =$

(16) $28 \div 7 =$ (17) $56 \div 8 =$ (18) $63 \div 7 =$

(19) $96 \div 12 =$ (20) $10 \div 2 =$ (21) $5 \div 5 =$

(22) $108 \div 12 =$ (23) $24 \div 8 =$ (24) $36 \div 3 =$

(25) $33 \div 3 =$ (26) $15 \div 5 =$ (27) $30 \div 3 =$

(28) $40 \div 4 =$ (29) $132 \div 12 =$ (30) $1 \div 1 =$

(31) $70 \div 7 =$ (32) $60 \div 6 =$ (33) $16 \div 4 =$

(34) $18 \div 9 =$ (35) $36 \div 4 =$ (36) $45 \div 5 =$

(37) $55 \div 11 =$ (38) $40 \div 8 =$ (39) $9 \div 9 =$

(40) $30 \div 10 =$ (41) $15 \div 3 =$ (42) $16 \div 4 =$

(43) $77 \div 7 =$ (44) $36 \div 3 =$ (45) $64 \div 8 =$

(46) $96 \div 12 =$ (47) $36 \div 9 =$ (48) $42 \div 6 =$

(49) $99 \div 11 =$ (50) $81 \div 9 =$ (51) $54 \div 9 =$

(52) $80 \div 8 =$ (53) $55 \div 5 =$ (54) $5 \div 5 =$

(55) $66 \div 6 =$ (56) $7 \div 1 =$ (57) $66 \div 11 =$

(58) $84 \div 12 =$ (59) $99 \div 9 =$ (60) $10 \div 5 =$

(1) 3 ÷ 1 =

(2) 88 ÷ 11 =

(3) 6 ÷ 6 =

(4) 144 ÷ 12 =

(5) 36 ÷ 3 =

(6) 12 ÷ 3 =

(7) 24 ÷ 2 =

(8) 63 ÷ 9 =

(9) 40 ÷ 8 =

(10) 88 ÷ 11 =

(11) 90 ÷ 10 =

(12) 18 ÷ 3 =

(13) 56 ÷ 7 =

(14) 35 ÷ 7 =

(15) 40 ÷ 10 =

(16) 30 ÷ 10 =

(17) 88 ÷ 8 =

(18) 121 ÷ 11 =

(19) 80 ÷ 8 =

(20) 7 ÷ 1 =

(21) 24 ÷ 12 =

(22) 3 ÷ 3 =

(23) 72 ÷ 6 =

(24) 72 ÷ 8 =

(25) 22 ÷ 11 =

(26) 63 ÷ 7 =

(27) 15 ÷ 3 =

(28) 70 ÷ 7 =

(29) 30 ÷ 10 =

(30) 18 ÷ 9 =

(31) 12 ÷ 4 =

(32) 10 ÷ 1 =

(33) 30 ÷ 3 =

(34) 22 ÷ 2 =

(35) 36 ÷ 9 =

(36) 72 ÷ 6 =

(37) 110 ÷ 11 =

(38) 3 ÷ 3 =

(39) 4 ÷ 4 =

(40) 4 ÷ 1 =

(41) 24 ÷ 12 =

(42) 70 ÷ 7 =

(43) 50 ÷ 10 =

(44) 8 ÷ 2 =

(45) 11 ÷ 11 =

(46) 9 ÷ 1 =

(47) 7 ÷ 7 =

(48) 64 ÷ 8 =

(49) 21 ÷ 3 =

(50) 27 ÷ 3 =

(51) 33 ÷ 3 =

(52) 80 ÷ 10 =

(53) 33 ÷ 11 =

(54) 12 ÷ 1 =

(55) 50 ÷ 5 =

(56) 28 ÷ 7 =

(57) 36 ÷ 9 =

(58) 12 ÷ 2 =

(59) 42 ÷ 6 =

(60) 9 ÷ 1 =

(1) 24 ÷ 12 =

(2) 56 ÷ 7 =

(3) 77 ÷ 11 =

(4) 25 ÷ 5 =

(5) 30 ÷ 5 =

(6) 24 ÷ 3 =

(7) 35 ÷ 5 =

(8) 2 ÷ 2 =

(9) 36 ÷ 12 =

(10) 40 ÷ 4 =

(11) 48 ÷ 12 =

(12) 11 ÷ 11 =

(13) 2 ÷ 1 =

(14) 72 ÷ 6 =

(15) 27 ÷ 9 =

(16) 35 ÷ 7 =

(17) 3 ÷ 3 =

(18) 15 ÷ 3 =

(19) 66 ÷ 6 =

(20) 14 ÷ 7 =

(21) 56 ÷ 8 =

(22) 44 ÷ 11 =

(23) 36 ÷ 6 =

(24) 20 ÷ 10 =

(25) 12 ÷ 1 =

(26) 8 ÷ 2 =

(27) 110 ÷ 11 =

(28) 48 ÷ 8 =

(29) 64 ÷ 8 =

(30) 4 ÷ 1 =

(31) 56 ÷ 8 =

(32) 90 ÷ 10 =

(33) 84 ÷ 12 =

(34) 4 ÷ 4 =

(35) 24 ÷ 12 =

(36) 40 ÷ 5 =

(37) 84 ÷ 12 =

(38) 28 ÷ 4 =

(39) 21 ÷ 3 =

(40) 24 ÷ 6 =

(41) 12 ÷ 1 =

(42) 4 ÷ 1 =

(43) 70 ÷ 10 =

(44) 12 ÷ 1 =

(45) 2 ÷ 2 =

(46) 18 ÷ 3 =

(47) 96 ÷ 12 =

(48) 8 ÷ 2 =

(49) 72 ÷ 9 =

(50) 110 ÷ 10 =

(51) 42 ÷ 7 =

(52) 72 ÷ 6 =

(53) 72 ÷ 8 =

(54) 60 ÷ 5 =

(55) 8 ÷ 1 =

(56) 5 ÷ 1 =

(57) 30 ÷ 3 =

(58) 84 ÷ 7 =

(59) 42 ÷ 6 =

(60) 35 ÷ 5 =

(1) $96 \div 8 =$

(2) $16 \div 4 =$

(3) $55 \div 5 =$

(4) $32 \div 8 =$

(5) $3 \div 3 =$

(6) $9 \div 1 =$

(7) $8 \div 4 =$

(8) $3 \div 3 =$

(9) $40 \div 8 =$

(10) $33 \div 11 =$

(11) $56 \div 8 =$

(12) $24 \div 6 =$

(13) $88 \div 8 =$

(14) $2 \div 1 =$

(15) $18 \div 9 =$

(16) $55 \div 11 =$

(17) $56 \div 7 =$

(18) $44 \div 4 =$

(19) $63 \div 7 =$

(20) $60 \div 5 =$

(21) $21 \div 7 =$

(22) $96 \div 12 =$

(23) $12 \div 3 =$

(24) $44 \div 4 =$

(25) $44 \div 11 =$

(26) $9 \div 9 =$

(27) $132 \div 11 =$

(28) $24 \div 12 =$

(29) $100 \div 10 =$

(30) $10 \div 5 =$

(31) $60 \div 12 =$

(32) $100 \div 10 =$

(33) $45 \div 9 =$

(34) $1 \div 1 =$

(35) $50 \div 10 =$

(36) $24 \div 6 =$

(37) $48 \div 12 =$

(38) $110 \div 10 =$

(39) $8 \div 1 =$

(40) $6 \div 2 =$

(41) $48 \div 6 =$

(42) $15 \div 3 =$

(43) $24 \div 3 =$

(44) $81 \div 9 =$

(45) $96 \div 8 =$

(46) $99 \div 11 =$

(47) $12 \div 12 =$

(48) $1 \div 1 =$

(49) $8 \div 2 =$

(50) $24 \div 6 =$

(51) $2 \div 1 =$

(52) $24 \div 2 =$

(53) $70 \div 7 =$

(54) $14 \div 7 =$

(55) $96 \div 12 =$

(56) $48 \div 12 =$

(57) $22 \div 11 =$

(58) $32 \div 4 =$

(59) $88 \div 8 =$

(60) $16 \div 8 =$

(1) $99 \div 11 =$

(2) $120 \div 12 =$

(3) $60 \div 10 =$

(4) $24 \div 8 =$

(5) $18 \div 2 =$

(6) $36 \div 9 =$

(7) $44 \div 11 =$

(8) $36 \div 4 =$

(9) $45 \div 5 =$

(10) $18 \div 3 =$

(11) $49 \div 7 =$

(12) $16 \div 2 =$

(13) $3 \div 1 =$

(14) $32 \div 8 =$

(15) $77 \div 7 =$

(16) $12 \div 1 =$

(17) $60 \div 5 =$

(18) $12 \div 6 =$

(19) $96 \div 8 =$

(20) $21 \div 3 =$

(21) $11 \div 1 =$

(22) $90 \div 9 =$

(23) $3 \div 3 =$

(24) $42 \div 6 =$

(25) $50 \div 10 =$

(26) $9 \div 1 =$

(27) $70 \div 7 =$

(28) $99 \div 9 =$

(29) $49 \div 7 =$

(30) $63 \div 9 =$

(31) $132 \div 11 =$

(32) $99 \div 9 =$

(33) $6 \div 3 =$

(34) $30 \div 5 =$

(35) $21 \div 7 =$

(36) $36 \div 4 =$

(37) $40 \div 10 =$

(38) $64 \div 8 =$

(39) $24 \div 12 =$

(40) $24 \div 3 =$

(41) $8 \div 4 =$

(42) $120 \div 12 =$

(43) $6 \div 6 =$

(44) $24 \div 6 =$

(45) $48 \div 8 =$

(46) $30 \div 3 =$

(47) $15 \div 5 =$

(48) $90 \div 10 =$

(49) $7 \div 1 =$

(50) $54 \div 6 =$

(51) $22 \div 11 =$

(52) $77 \div 11 =$

(53) $55 \div 11 =$

(54) $20 \div 2 =$

(55) $70 \div 10 =$

(56) $84 \div 7 =$

(57) $7 \div 7 =$

(58) $16 \div 4 =$

(59) $88 \div 8 =$

(60) $36 \div 4 =$

(1) $72 \div 8 =$

(2) $18 \div 2 =$

(3) $48 \div 12 =$

(4) $60 \div 5 =$

(5) $7 \div 1 =$

(6) $36 \div 12 =$

(7) $30 \div 5 =$

(8) $4 \div 4 =$

(9) $100 \div 10 =$

(10) $50 \div 10 =$

(11) $33 \div 11 =$

(12) $36 \div 4 =$

(13) $12 \div 3 =$

(14) $144 \div 12 =$

(15) $11 \div 1 =$

(16) $24 \div 3 =$

(17) $24 \div 6 =$

(18) $24 \div 8 =$

(19) $80 \div 10 =$

(20) $4 \div 1 =$

(21) $24 \div 8 =$

(22) $70 \div 7 =$

(23) $1 \div 1 =$

(24) $30 \div 5 =$

(25) $99 \div 9 =$

(26) $10 \div 2 =$

(27) $36 \div 4 =$

(28) $90 \div 10 =$

(29) $18 \div 9 =$

(30) $66 \div 6 =$

(31) $30 \div 6 =$

(32) $80 \div 8 =$

(33) $5 \div 5 =$

(34) $56 \div 8 =$

(35) $4 \div 1 =$

(36) $35 \div 5 =$

(37) $4 \div 4 =$

(38) $24 \div 12 =$

(39) $12 \div 2 =$

(40) $12 \div 4 =$

(41) $80 \div 10 =$

(42) $144 \div 12 =$

(43) $30 \div 10 =$

(44) $15 \div 5 =$

(45) $63 \div 7 =$

(46) $108 \div 9 =$

(47) $12 \div 4 =$

(48) $110 \div 11 =$

(49) $60 \div 12 =$

(50) $81 \div 9 =$

(51) $54 \div 6 =$

(52) $12 \div 6 =$

(53) $108 \div 12 =$

(54) $35 \div 5 =$

(55) $4 \div 1 =$

(56) $48 \div 8 =$

(57) $110 \div 10 =$

(58) $84 \div 12 =$

(59) $60 \div 10 =$

(60) $14 \div 2 =$

(1) 56 ÷ 8 =

(2) 42 ÷ 7 =

(3) 14 ÷ 2 =

(4) 12 ÷ 2 =

(5) 32 ÷ 4 =

(6) 27 ÷ 3 =

(7) 60 ÷ 5 =

(8) 15 ÷ 3 =

(9) 5 ÷ 1 =

(10) 27 ÷ 3 =

(11) 88 ÷ 11 =

(12) 40 ÷ 8 =

(13) 7 ÷ 1 =

(14) 20 ÷ 2 =

(15) 35 ÷ 5 =

(16) 72 ÷ 9 =

(17) 12 ÷ 6 =

(18) 30 ÷ 6 =

(19) 2 ÷ 2 =

(20) 5 ÷ 1 =

(21) 60 ÷ 10 =

(22) 24 ÷ 4 =

(23) 33 ÷ 3 =

(24) 18 ÷ 6 =

(25) 4 ÷ 4 =

(26) 8 ÷ 8 =

(27) 60 ÷ 10 =

(28) 6 ÷ 1 =

(29) 12 ÷ 1 =

(30) 60 ÷ 6 =

(31) 72 ÷ 9 =

(32) 14 ÷ 7 =

(33) 121 ÷ 11 =

(34) 12 ÷ 3 =

(35) 72 ÷ 12 =

(36) 32 ÷ 8 =

(37) 10 ÷ 10 =

(38) 30 ÷ 6 =

(39) 16 ÷ 2 =

(40) 2 ÷ 2 =

(41) 33 ÷ 3 =

(42) 42 ÷ 6 =

(43) 48 ÷ 4 =

(44) 16 ÷ 4 =

(45) 45 ÷ 9 =

(46) 25 ÷ 5 =

(47) 70 ÷ 10 =

(48) 14 ÷ 2 =

(49) 30 ÷ 3 =

(50) 27 ÷ 3 =

(51) 110 ÷ 11 =

(52) 10 ÷ 10 =

(53) 14 ÷ 2 =

(54) 72 ÷ 6 =

(55) 15 ÷ 3 =

(56) 54 ÷ 6 =

(57) 36 ÷ 3 =

(58) 4 ÷ 1 =

(59) 18 ÷ 6 =

(60) 35 ÷ 5 =

(1) $64 \div 8 =$

(2) $3 \div 1 =$

(3) $132 \div 12 =$

(4) $15 \div 5 =$

(5) $18 \div 6 =$

(6) $2 \div 2 =$

(7) $90 \div 10 =$

(8) $54 \div 9 =$

(9) $80 \div 10 =$

(10) $18 \div 6 =$

(11) $30 \div 5 =$

(12) $60 \div 5 =$

(13) $30 \div 5 =$

(14) $54 \div 9 =$

(15) $20 \div 4 =$

(16) $35 \div 5 =$

(17) $144 \div 12 =$

(18) $70 \div 10 =$

(19) $4 \div 2 =$

(20) $8 \div 1 =$

(21) $33 \div 3 =$

(22) $22 \div 11 =$

(23) $60 \div 12 =$

(24) $18 \div 3 =$

(25) $48 \div 12 =$

(26) $8 \div 2 =$

(27) $64 \div 8 =$

(28) $72 \div 8 =$

(29) $121 \div 11 =$

(30) $24 \div 8 =$

(31) $3 \div 3 =$

(32) $36 \div 4 =$

(33) $24 \div 4 =$

(34) $30 \div 6 =$

(35) $121 \div 11 =$

(36) $80 \div 8 =$

(37) $32 \div 4 =$

(38) $48 \div 8 =$

(39) $33 \div 11 =$

(40) $66 \div 11 =$

(41) $60 \div 6 =$

(42) $132 \div 11 =$

(43) $4 \div 2 =$

(44) $72 \div 9 =$

(45) $2 \div 2 =$

(46) $84 \div 7 =$

(47) $45 \div 9 =$

(48) $96 \div 8 =$

(49) $36 \div 6 =$

(50) $66 \div 11 =$

(51) $30 \div 3 =$

(52) $3 \div 3 =$

(53) $6 \div 2 =$

(54) $10 \div 2 =$

(55) $4 \div 2 =$

(56) $44 \div 11 =$

(57) $30 \div 10 =$

(58) $12 \div 6 =$

(59) $8 \div 2 =$

(60) $108 \div 9 =$

(1) 10 ÷ 2 =

(2) 24 ÷ 12 =

(3) 15 ÷ 5 =

(4) 5 ÷ 1 =

(5) 6 ÷ 6 =

(6) 44 ÷ 4 =

(7) 63 ÷ 9 =

(8) 12 ÷ 1 =

(9) 11 ÷ 1 =

(10) 70 ÷ 10 =

(11) 70 ÷ 7 =

(12) 81 ÷ 9 =

(13) 7 ÷ 7 =

(14) 14 ÷ 7 =

(15) 48 ÷ 6 =

(16) 70 ÷ 10 =

(17) 6 ÷ 2 =

(18) 42 ÷ 7 =

(19) 110 ÷ 11 =

(20) 70 ÷ 7 =

(21) 8 ÷ 2 =

(22) 22 ÷ 2 =

(23) 12 ÷ 12 =

(24) 48 ÷ 6 =

(25) 12 ÷ 6 =

(26) 28 ÷ 4 =

(27) 7 ÷ 7 =

(28) 40 ÷ 5 =

(29) 1 ÷ 1 =

(30) 6 ÷ 1 =

(31) 28 ÷ 4 =

(32) 48 ÷ 6 =

(33) 40 ÷ 4 =

(34) 27 ÷ 9 =

(35) 132 ÷ 12 =

(36) 22 ÷ 2 =

(37) 96 ÷ 12 =

(38) 99 ÷ 9 =

(39) 80 ÷ 8 =

(40) 8 ÷ 2 =

(41) 44 ÷ 4 =

(42) 6 ÷ 3 =

(43) 12 ÷ 4 =

(44) 7 ÷ 1 =

(45) 77 ÷ 11 =

(46) 24 ÷ 6 =

(47) 96 ÷ 12 =

(48) 1 ÷ 1 =

(49) 6 ÷ 6 =

(50) 11 ÷ 1 =

(51) 70 ÷ 10 =

(52) 66 ÷ 11 =

(53) 18 ÷ 9 =

(54) 24 ÷ 12 =

(55) 110 ÷ 11 =

(56) 88 ÷ 11 =

(57) 72 ÷ 8 =

(58) 54 ÷ 9 =

(59) 7 ÷ 7 =

(60) 24 ÷ 2 =

(1) $36 \div 4 =$

(2) $18 \div 6 =$

(3) $80 \div 10 =$

(4) $110 \div 10 =$

(5) $11 \div 1 =$

(6) $32 \div 4 =$

(7) $18 \div 9 =$

(8) $110 \div 11 =$

(9) $18 \div 9 =$

(10) $5 \div 5 =$

(11) $50 \div 10 =$

(12) $60 \div 12 =$

(13) $48 \div 8 =$

(14) $108 \div 12 =$

(15) $30 \div 6 =$

(16) $12 \div 6 =$

(17) $100 \div 10 =$

(18) $40 \div 10 =$

(19) $27 \div 9 =$

(20) $45 \div 5 =$

(21) $30 \div 10 =$

(22) $144 \div 12 =$

(23) $64 \div 8 =$

(24) $4 \div 4 =$

(25) $15 \div 3 =$

(26) $12 \div 12 =$

(27) $4 \div 4 =$

(28) $35 \div 7 =$

(29) $5 \div 1 =$

(30) $4 \div 2 =$

(31) $32 \div 4 =$

(32) $4 \div 4 =$

(33) $90 \div 9 =$

(34) $14 \div 2 =$

(35) $56 \div 8 =$

(36) $10 \div 2 =$

(37) $108 \div 9 =$

(38) $27 \div 3 =$

(39) $4 \div 4 =$

(40) $35 \div 7 =$

(41) $110 \div 10 =$

(42) $36 \div 3 =$

(43) $110 \div 11 =$

(44) $96 \div 12 =$

(45) $77 \div 11 =$

(46) $40 \div 8 =$

(47) $20 \div 10 =$

(48) $54 \div 6 =$

(49) $55 \div 11 =$

(50) $120 \div 12 =$

(51) $9 \div 1 =$

(52) $16 \div 8 =$

(53) $108 \div 9 =$

(54) $5 \div 5 =$

(55) $48 \div 12 =$

(56) $12 \div 4 =$

(57) $36 \div 6 =$

(58) $108 \div 9 =$

(59) $32 \div 8 =$

(60) $30 \div 5 =$

(1) $22 \div 11 =$

(2) $44 \div 11 =$

(3) $96 \div 12 =$

(4) $30 \div 3 =$

(5) $11 \div 1 =$

(6) $2 \div 1 =$

(7) $11 \div 11 =$

(8) $20 \div 5 =$

(9) $84 \div 7 =$

(10) $22 \div 11 =$

(11) $132 \div 12 =$

(12) $132 \div 11 =$

(13) $45 \div 5 =$

(14) $12 \div 2 =$

(15) $14 \div 7 =$

(16) $96 \div 12 =$

(17) $90 \div 9 =$

(18) $30 \div 6 =$

(19) $45 \div 5 =$

(20) $60 \div 6 =$

(21) $56 \div 8 =$

(22) $44 \div 11 =$

(23) $6 \div 2 =$

(24) $4 \div 1 =$

(25) $24 \div 12 =$

(26) $84 \div 7 =$

(27) $33 \div 3 =$

(28) $80 \div 10 =$

(29) $90 \div 10 =$

(30) $70 \div 7 =$

(31) $28 \div 7 =$

(32) $66 \div 11 =$

(33) $6 \div 3 =$

(34) $28 \div 4 =$

(35) $8 \div 4 =$

(36) $56 \div 7 =$

(37) $28 \div 4 =$

(38) $48 \div 6 =$

(39) $64 \div 8 =$

(40) $50 \div 10 =$

(41) $16 \div 8 =$

(42) $49 \div 7 =$

(43) $63 \div 7 =$

(44) $121 \div 11 =$

(45) $36 \div 4 =$

(46) $108 \div 9 =$

(47) $72 \div 12 =$

(48) $25 \div 5 =$

(49) $60 \div 12 =$

(50) $60 \div 6 =$

(51) $49 \div 7 =$

(52) $60 \div 5 =$

(53) $27 \div 9 =$

(54) $24 \div 2 =$

(55) $5 \div 1 =$

(56) $24 \div 6 =$

(57) $20 \div 5 =$

(58) $40 \div 8 =$

(59) $42 \div 7 =$

(60) $25 \div 5 =$

| Day: | 90 | Date: | | Score: | /60 |
| Name: | | Time: | : | Rating: | ☆☆☆☆☆ |

(1) $77 \div 11 =$

(2) $54 \div 9 =$

(3) $96 \div 12 =$

(4) $63 \div 9 =$

(5) $60 \div 6 =$

(6) $9 \div 3 =$

(7) $144 \div 12 =$

(8) $50 \div 10 =$

(9) $36 \div 4 =$

(10) $20 \div 5 =$

(11) $84 \div 7 =$

(12) $66 \div 6 =$

(13) $12 \div 1 =$

(14) $20 \div 5 =$

(15) $120 \div 10 =$

(16) $72 \div 8 =$

(17) $9 \div 3 =$

(18) $40 \div 8 =$

(19) $36 \div 3 =$

(20) $28 \div 4 =$

(21) $99 \div 11 =$

(22) $6 \div 2 =$

(23) $36 \div 3 =$

(24) $9 \div 9 =$

(25) $25 \div 5 =$

(26) $110 \div 10 =$

(27) $5 \div 5 =$

(28) $2 \div 1 =$

(29) $18 \div 2 =$

(30) $121 \div 11 =$

(31) $8 \div 8 =$

(32) $88 \div 11 =$

(33) $9 \div 9 =$

(34) $81 \div 9 =$

(35) $30 \div 10 =$

(36) $30 \div 5 =$

(37) $90 \div 10 =$

(38) $10 \div 1 =$

(39) $12 \div 12 =$

(40) $8 \div 4 =$

(41) $20 \div 10 =$

(42) $42 \div 6 =$

(43) $40 \div 4 =$

(44) $11 \div 11 =$

(45) $32 \div 4 =$

(46) $4 \div 2 =$

(47) $49 \div 7 =$

(48) $66 \div 11 =$

(49) $6 \div 2 =$

(50) $22 \div 11 =$

(51) $3 \div 1 =$

(52) $2 \div 1 =$

(53) $132 \div 11 =$

(54) $14 \div 7 =$

(55) $10 \div 2 =$

(56) $32 \div 4 =$

(57) $33 \div 3 =$

(58) $1 \div 1 =$

(59) $16 \div 8 =$

(60) $12 \div 4 =$

(1) $6 \div 1 =$

(2) $48 \div 6 =$

(3) $24 \div 4 =$

(4) $12 \div 1 =$

(5) $25 \div 5 =$

(6) $80 \div 8 =$

(7) $72 \div 6 =$

(8) $42 \div 6 =$

(9) $96 \div 12 =$

(10) $8 \div 4 =$

(11) $36 \div 6 =$

(12) $55 \div 11 =$

(13) $24 \div 4 =$

(14) $32 \div 8 =$

(15) $60 \div 6 =$

(16) $1 \div 1 =$

(17) $33 \div 11 =$

(18) $77 \div 11 =$

(19) $30 \div 6 =$

(20) $72 \div 8 =$

(21) $55 \div 11 =$

(22) $6 \div 2 =$

(23) $80 \div 10 =$

(24) $4 \div 4 =$

(25) $5 \div 1 =$

(26) $90 \div 9 =$

(27) $28 \div 7 =$

(28) $12 \div 1 =$

(29) $20 \div 2 =$

(30) $36 \div 6 =$

(31) $12 \div 4 =$

(32) $6 \div 1 =$

(33) $18 \div 2 =$

(34) $40 \div 10 =$

(35) $77 \div 7 =$

(36) $2 \div 2 =$

(37) $72 \div 9 =$

(38) $10 \div 1 =$

(39) $110 \div 10 =$

(40) $10 \div 10 =$

(41) $9 \div 9 =$

(42) $99 \div 11 =$

(43) $50 \div 10 =$

(44) $5 \div 5 =$

(45) $9 \div 3 =$

(46) $12 \div 4 =$

(47) $6 \div 1 =$

(48) $6 \div 6 =$

(49) $8 \div 4 =$

(50) $60 \div 10 =$

(51) $80 \div 10 =$

(52) $18 \div 3 =$

(53) $12 \div 4 =$

(54) $28 \div 7 =$

(55) $48 \div 4 =$

(56) $36 \div 12 =$

(57) $84 \div 12 =$

(58) $56 \div 7 =$

(59) $30 \div 3 =$

(60) $5 \div 5 =$

Day: 92   Date: Score: /60

Name: Time: :   Rating: ☆☆☆☆☆

(1) $8 \div 8 =$

(2) $9 \div 9 =$

(3) $70 \div 7 =$

(4) $3 \div 3 =$

(5) $99 \div 11 =$

(6) $40 \div 4 =$

(7) $16 \div 2 =$

(8) $20 \div 5 =$

(9) $24 \div 3 =$

(10) $96 \div 8 =$

(11) $50 \div 5 =$

(12) $25 \div 5 =$

(13) $77 \div 7 =$

(14) $28 \div 4 =$

(15) $110 \div 11 =$

(16) $12 \div 3 =$

(17) $24 \div 8 =$

(18) $14 \div 7 =$

(19) $48 \div 6 =$

(20) $6 \div 2 =$

(21) $40 \div 5 =$

(22) $32 \div 4 =$

(23) $99 \div 9 =$

(24) $8 \div 1 =$

(25) $55 \div 11 =$

(26) $40 \div 10 =$

(27) $9 \div 9 =$

(28) $66 \div 11 =$

(29) $10 \div 10 =$

(30) $63 \div 7 =$

(31) $18 \div 9 =$

(32) $56 \div 7 =$

(33) $120 \div 10 =$

(34) $42 \div 7 =$

(35) $35 \div 5 =$

(36) $20 \div 4 =$

(37) $90 \div 9 =$

(38) $4 \div 2 =$

(39) $33 \div 11 =$

(40) $4 \div 2 =$

(41) $70 \div 10 =$

(42) $77 \div 7 =$

(43) $27 \div 9 =$

(44) $28 \div 7 =$

(45) $60 \div 12 =$

(46) $60 \div 12 =$

(47) $35 \div 5 =$

(48) $8 \div 1 =$

(49) $6 \div 1 =$

(50) $6 \div 2 =$

(51) $110 \div 11 =$

(52) $8 \div 4 =$

(53) $16 \div 4 =$

(54) $35 \div 7 =$

(55) $33 \div 3 =$

(56) $6 \div 3 =$

(57) $25 \div 5 =$

(58) $120 \div 10 =$

(59) $22 \div 2 =$

(60) $72 \div 12 =$

(1) $25 \div 5 =$

(2) $96 \div 8 =$

(3) $32 \div 8 =$

(4) $5 \div 5 =$

(5) $72 \div 12 =$

(6) $16 \div 2 =$

(7) $3 \div 1 =$

(8) $20 \div 10 =$

(9) $44 \div 11 =$

(10) $40 \div 10 =$

(11) $99 \div 11 =$

(12) $16 \div 8 =$

(13) $84 \div 12 =$

(14) $66 \div 11 =$

(15) $28 \div 7 =$

(16) $16 \div 4 =$

(17) $9 \div 3 =$

(18) $16 \div 4 =$

(19) $55 \div 5 =$

(20) $12 \div 1 =$

(21) $16 \div 8 =$

(22) $24 \div 4 =$

(23) $72 \div 9 =$

(24) $33 \div 11 =$

(25) $18 \div 6 =$

(26) $9 \div 9 =$

(27) $48 \div 4 =$

(28) $50 \div 10 =$

(29) $20 \div 2 =$

(30) $120 \div 12 =$

(31) $60 \div 5 =$

(32) $9 \div 1 =$

(33) $24 \div 3 =$

(34) $44 \div 4 =$

(35) $18 \div 2 =$

(36) $36 \div 6 =$

(37) $8 \div 8 =$

(38) $22 \div 2 =$

(39) $44 \div 11 =$

(40) $18 \div 2 =$

(41) $24 \div 4 =$

(42) $50 \div 5 =$

(43) $2 \div 2 =$

(44) $6 \div 6 =$

(45) $5 \div 5 =$

(46) $20 \div 10 =$

(47) $18 \div 3 =$

(48) $72 \div 8 =$

(49) $5 \div 1 =$

(50) $80 \div 10 =$

(51) $54 \div 6 =$

(52) $21 \div 3 =$

(53) $40 \div 10 =$

(54) $12 \div 2 =$

(55) $42 \div 6 =$

(56) $20 \div 4 =$

(57) $77 \div 11 =$

(58) $4 \div 2 =$

(59) $44 \div 11 =$

(60) $7 \div 1 =$

(1) $8 \div 1 =$

(2) $50 \div 10 =$

(3) $7 \div 1 =$

(4) $88 \div 8 =$

(5) $1 \div 1 =$

(6) $48 \div 12 =$

(7) $96 \div 8 =$

(8) $18 \div 9 =$

(9) $27 \div 3 =$

(10) $40 \div 4 =$

(11) $54 \div 6 =$

(12) $72 \div 6 =$

(13) $18 \div 9 =$

(14) $40 \div 8 =$

(15) $24 \div 3 =$

(16) $20 \div 2 =$

(17) $84 \div 12 =$

(18) $54 \div 6 =$

(19) $12 \div 6 =$

(20) $35 \div 7 =$

(21) $40 \div 8 =$

(22) $40 \div 5 =$

(23) $18 \div 3 =$

(24) $66 \div 6 =$

(25) $55 \div 11 =$

(26) $80 \div 8 =$

(27) $24 \div 3 =$

(28) $12 \div 12 =$

(29) $21 \div 3 =$

(30) $120 \div 12 =$

(31) $110 \div 10 =$

(32) $80 \div 10 =$

(33) $24 \div 12 =$

(34) $54 \div 6 =$

(35) $56 \div 8 =$

(36) $9 \div 1 =$

(37) $10 \div 10 =$

(38) $18 \div 6 =$

(39) $36 \div 4 =$

(40) $88 \div 11 =$

(41) $20 \div 10 =$

(42) $12 \div 2 =$

(43) $48 \div 4 =$

(44) $44 \div 11 =$

(45) $81 \div 9 =$

(46) $80 \div 8 =$

(47) $88 \div 8 =$

(48) $48 \div 6 =$

(49) $27 \div 3 =$

(50) $14 \div 7 =$

(51) $9 \div 1 =$

(52) $48 \div 6 =$

(53) $96 \div 8 =$

(54) $12 \div 4 =$

(55) $20 \div 4 =$

(56) $20 \div 10 =$

(57) $12 \div 6 =$

(58) $132 \div 11 =$

(59) $21 \div 3 =$

(60) $72 \div 12 =$

(1) $20 \div 4 =$

(2) $8 \div 1 =$

(3) $66 \div 6 =$

(4) $40 \div 8 =$

(5) $99 \div 11 =$

(6) $10 \div 2 =$

(7) $45 \div 5 =$

(8) $12 \div 6 =$

(9) $18 \div 6 =$

(10) $40 \div 4 =$

(11) $48 \div 4 =$

(12) $48 \div 8 =$

(13) $6 \div 3 =$

(14) $49 \div 7 =$

(15) $15 \div 3 =$

(16) $36 \div 3 =$

(17) $120 \div 12 =$

(18) $6 \div 3 =$

(19) $72 \div 8 =$

(20) $42 \div 7 =$

(21) $21 \div 3 =$

(22) $24 \div 8 =$

(23) $5 \div 1 =$

(24) $36 \div 12 =$

(25) $10 \div 5 =$

(26) $20 \div 2 =$

(27) $2 \div 1 =$

(28) $132 \div 12 =$

(29) $84 \div 7 =$

(30) $36 \div 4 =$

(31) $40 \div 4 =$

(32) $7 \div 7 =$

(33) $48 \div 8 =$

(34) $60 \div 5 =$

(35) $8 \div 2 =$

(36) $21 \div 3 =$

(37) $30 \div 5 =$

(38) $77 \div 11 =$

(39) $9 \div 9 =$

(40) $40 \div 8 =$

(41) $144 \div 12 =$

(42) $63 \div 9 =$

(43) $66 \div 11 =$

(44) $110 \div 10 =$

(45) $72 \div 9 =$

(46) $9 \div 9 =$

(47) $4 \div 1 =$

(48) $24 \div 2 =$

(49) $10 \div 5 =$

(50) $5 \div 5 =$

(51) $40 \div 4 =$

(52) $14 \div 7 =$

(53) $36 \div 9 =$

(54) $96 \div 8 =$

(55) $72 \div 9 =$

(56) $9 \div 1 =$

(57) $28 \div 4 =$

(58) $100 \div 10 =$

(59) $84 \div 12 =$

(60) $77 \div 11 =$

(1) $100 \div 10 =$

(2) $44 \div 4 =$

(3) $84 \div 7 =$

(4) $70 \div 10 =$

(5) $36 \div 3 =$

(6) $3 \div 1 =$

(7) $4 \div 1 =$

(8) $45 \div 9 =$

(9) $50 \div 10 =$

(10) $14 \div 2 =$

(11) $6 \div 6 =$

(12) $40 \div 4 =$

(13) $24 \div 2 =$

(14) $99 \div 9 =$

(15) $24 \div 8 =$

(16) $20 \div 4 =$

(17) $12 \div 12 =$

(18) $22 \div 11 =$

(19) $77 \div 11 =$

(20) $12 \div 12 =$

(21) $60 \div 12 =$

(22) $80 \div 8 =$

(23) $36 \div 12 =$

(24) $8 \div 2 =$

(25) $45 \div 9 =$

(26) $40 \div 5 =$

(27) $50 \div 5 =$

(28) $9 \div 3 =$

(29) $60 \div 5 =$

(30) $15 \div 5 =$

(31) $8 \div 4 =$

(32) $18 \div 2 =$

(33) $12 \div 6 =$

(34) $36 \div 12 =$

(35) $40 \div 4 =$

(36) $20 \div 10 =$

(37) $36 \div 12 =$

(38) $10 \div 1 =$

(39) $56 \div 8 =$

(40) $99 \div 11 =$

(41) $27 \div 9 =$

(42) $21 \div 7 =$

(43) $8 \div 1 =$

(44) $49 \div 7 =$

(45) $54 \div 6 =$

(46) $3 \div 1 =$

(47) $84 \div 12 =$

(48) $3 \div 1 =$

(49) $63 \div 7 =$

(50) $144 \div 12 =$

(51) $63 \div 7 =$

(52) $64 \div 8 =$

(53) $20 \div 5 =$

(54) $24 \div 8 =$

(55) $2 \div 1 =$

(56) $44 \div 11 =$

(57) $72 \div 6 =$

(58) $90 \div 9 =$

(59) $2 \div 1 =$

(60) $60 \div 6 =$

(1) $80 \div 10 =$

(2) $45 \div 9 =$

(3) $15 \div 5 =$

(4) $70 \div 7 =$

(5) $20 \div 2 =$

(6) $2 \div 2 =$

(7) $10 \div 2 =$

(8) $24 \div 3 =$

(9) $77 \div 11 =$

(10) $49 \div 7 =$

(11) $25 \div 5 =$

(12) $16 \div 4 =$

(13) $54 \div 9 =$

(14) $33 \div 11 =$

(15) $21 \div 3 =$

(16) $12 \div 4 =$

(17) $60 \div 12 =$

(18) $28 \div 7 =$

(19) $30 \div 5 =$

(20) $3 \div 1 =$

(21) $24 \div 6 =$

(22) $28 \div 4 =$

(23) $6 \div 3 =$

(24) $96 \div 8 =$

(25) $35 \div 5 =$

(26) $120 \div 10 =$

(27) $25 \div 5 =$

(28) $55 \div 5 =$

(29) $21 \div 7 =$

(30) $36 \div 4 =$

(31) $63 \div 9 =$

(32) $11 \div 1 =$

(33) $45 \div 5 =$

(34) $30 \div 3 =$

(35) $45 \div 9 =$

(36) $40 \div 4 =$

(37) $60 \div 5 =$

(38) $90 \div 9 =$

(39) $12 \div 2 =$

(40) $88 \div 8 =$

(41) $40 \div 10 =$

(42) $96 \div 12 =$

(43) $24 \div 2 =$

(44) $42 \div 6 =$

(45) $12 \div 3 =$

(46) $81 \div 9 =$

(47) $121 \div 11 =$

(48) $30 \div 10 =$

(49) $9 \div 9 =$

(50) $60 \div 12 =$

(51) $18 \div 3 =$

(52) $144 \div 12 =$

(53) $70 \div 10 =$

(54) $108 \div 9 =$

(55) $2 \div 1 =$

(56) $14 \div 7 =$

(57) $8 \div 8 =$

(58) $12 \div 12 =$

(59) $27 \div 3 =$

(60) $14 \div 2 =$

(1) 18 ÷ 2 =

(2) 33 ÷ 3 =

(3) 1 ÷ 1 =

(4) 20 ÷ 5 =

(5) 80 ÷ 8 =

(6) 12 ÷ 4 =

(7) 48 ÷ 6 =

(8) 32 ÷ 4 =

(9) 15 ÷ 5 =

(10) 44 ÷ 4 =

(11) 20 ÷ 5 =

(12) 54 ÷ 6 =

(13) 63 ÷ 9 =

(14) 88 ÷ 8 =

(15) 14 ÷ 7 =

(16) 40 ÷ 4 =

(17) 20 ÷ 10 =

(18) 20 ÷ 4 =

(19) 60 ÷ 6 =

(20) 12 ÷ 4 =

(21) 18 ÷ 3 =

(22) 5 ÷ 5 =

(23) 90 ÷ 10 =

(24) 72 ÷ 9 =

(25) 3 ÷ 3 =

(26) 36 ÷ 6 =

(27) 40 ÷ 10 =

(28) 84 ÷ 12 =

(29) 35 ÷ 5 =

(30) 27 ÷ 9 =

(31) 80 ÷ 10 =

(32) 88 ÷ 8 =

(33) 22 ÷ 2 =

(34) 132 ÷ 12 =

(35) 48 ÷ 4 =

(36) 7 ÷ 1 =

(37) 40 ÷ 8 =

(38) 6 ÷ 6 =

(39) 110 ÷ 11 =

(40) 60 ÷ 12 =

(41) 4 ÷ 2 =

(42) 22 ÷ 11 =

(43) 44 ÷ 4 =

(44) 108 ÷ 12 =

(45) 9 ÷ 1 =

(46) 72 ÷ 8 =

(47) 9 ÷ 1 =

(48) 77 ÷ 11 =

(49) 20 ÷ 5 =

(50) 21 ÷ 7 =

(51) 72 ÷ 8 =

(52) 45 ÷ 9 =

(53) 24 ÷ 3 =

(54) 48 ÷ 8 =

(55) 48 ÷ 8 =

(56) 36 ÷ 9 =

(57) 9 ÷ 3 =

(58) 55 ÷ 11 =

(59) 63 ÷ 9 =

(60) 77 ÷ 7 =

Day:	99	

Date:		Score:	/60
Name:		Time: :	Rating: ☆☆☆☆☆☆

(1) 40 ÷ 4 =

(2) 132 ÷ 11 =

(3) 63 ÷ 9 =

(4) 33 ÷ 11 =

(5) 30 ÷ 3 =

(6) 60 ÷ 6 =

(7) 4 ÷ 2 =

(8) 7 ÷ 1 =

(9) 120 ÷ 10 =

(10) 63 ÷ 7 =

(11) 11 ÷ 11 =

(12) 6 ÷ 1 =

(13) 99 ÷ 9 =

(14) 70 ÷ 10 =

(15) 60 ÷ 5 =

(16) 70 ÷ 7 =

(17) 18 ÷ 6 =

(18) 35 ÷ 5 =

(19) 24 ÷ 8 =

(20) 14 ÷ 7 =

(21) 24 ÷ 4 =

(22) 9 ÷ 3 =

(23) 3 ÷ 3 =

(24) 45 ÷ 9 =

(25) 56 ÷ 7 =

(26) 50 ÷ 5 =

(27) 28 ÷ 4 =

(28) 56 ÷ 7 =

(29) 70 ÷ 7 =

(30) 7 ÷ 1 =

(31) 40 ÷ 5 =

(32) 16 ÷ 4 =

(33) 5 ÷ 5 =

(34) 42 ÷ 7 =

(35) 20 ÷ 4 =

(36) 24 ÷ 12 =

(37) 22 ÷ 11 =

(38) 16 ÷ 8 =

(39) 3 ÷ 1 =

(40) 5 ÷ 1 =

(41) 10 ÷ 1 =

(42) 16 ÷ 2 =

(43) 81 ÷ 9 =

(44) 9 ÷ 9 =

(45) 42 ÷ 7 =

(46) 8 ÷ 1 =

(47) 132 ÷ 11 =

(48) 90 ÷ 10 =

(49) 81 ÷ 9 =

(50) 132 ÷ 12 =

(51) 24 ÷ 4 =

(52) 12 ÷ 1 =

(53) 60 ÷ 10 =

(54) 12 ÷ 12 =

(55) 12 ÷ 1 =

(56) 9 ÷ 1 =

(57) 10 ÷ 1 =

(58) 10 ÷ 1 =

(59) 10 ÷ 5 =

(60) 24 ÷ 4 =

Day: 100

Date:

Score: /60

Name:

Time: :

Rating: ☆☆☆☆☆☆

(1) 77 ÷ 7 =

(2) 16 ÷ 8 =

(3) 132 ÷ 11 =

(4) 18 ÷ 9 =

(5) 54 ÷ 9 =

(6) 11 ÷ 11 =

(7) 44 ÷ 11 =

(8) 20 ÷ 2 =

(9) 120 ÷ 10 =

(10) 3 ÷ 3 =

(11) 9 ÷ 1 =

(12) 100 ÷ 10 =

(13) 50 ÷ 10 =

(14) 10 ÷ 2 =

(15) 18 ÷ 3 =

(16) 18 ÷ 2 =

(17) 10 ÷ 1 =

(18) 132 ÷ 11 =

(19) 50 ÷ 5 =

(20) 32 ÷ 4 =

(21) 64 ÷ 8 =

(22) 56 ÷ 8 =

(23) 30 ÷ 3 =

(24) 27 ÷ 3 =

(25) 35 ÷ 7 =

(26) 18 ÷ 2 =

(27) 35 ÷ 5 =

(28) 16 ÷ 4 =

(29) 12 ÷ 2 =

(30) 27 ÷ 9 =

(31) 22 ÷ 11 =

(32) 100 ÷ 10 =

(33) 72 ÷ 6 =

(34) 110 ÷ 10 =

(35) 32 ÷ 4 =

(36) 10 ÷ 5 =

(37) 22 ÷ 2 =

(38) 15 ÷ 3 =

(39) 24 ÷ 6 =

(40) 12 ÷ 1 =

(41) 80 ÷ 8 =

(42) 33 ÷ 11 =

(43) 120 ÷ 10 =

(44) 16 ÷ 8 =

(45) 66 ÷ 11 =

(46) 90 ÷ 10 =

(47) 108 ÷ 12 =

(48) 3 ÷ 1 =

(49) 6 ÷ 2 =

(50) 8 ÷ 2 =

(51) 60 ÷ 5 =

(52) 120 ÷ 12 =

(53) 12 ÷ 3 =

(54) 16 ÷ 4 =

(55) 14 ÷ 7 =

(56) 18 ÷ 9 =

(57) 12 ÷ 1 =

(58) 12 ÷ 2 =

(59) 10 ÷ 10 =

(60) 27 ÷ 9 =

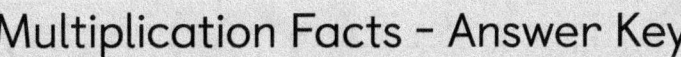

Multiplication Facts - Answer Key

DAY 1
1) 14 2) 24 3) 20
4) 20 5) 15 6) 16
7) 32 8) 12 9) 11
10) 70 11) 36 12) 72
13) 27 14) 30 15) 16
16) 42 17) 11 18) 8
19) 24 20) 40 21) 2
22) 5 23) 30 24) 4
25) 7 26) 84 27) 50
28) 3 29) 24 30) 6
31) 132 32) 72 33) 80
34) 3 35) 6 36) 88
37) 18 38) 132 39) 40
40) 90 41) 96 42) 6
43) 44 44) 99 45) 60
46) 30 47) 20 48) 36
49) 8 50) 20 51) 8
52) 12 53) 18 54) 25
55) 77 56) 110 57) 10
58) 72 59) 33 60) 84

DAY 2
1) 121 2) 15 3) 108
4) 11 5) 45 6) 10
7) 6 8) 6 9) 12
10) 28 11) 108 12) 88
13) 24 14) 56 15) 3
16) 3 17) 90 18) 54
19) 6 20) 55 21) 20
22) 10 23) 48 24) 8
25) 144 26) 15 27) 66
28) 66 29) 22 30) 10
31) 72 32) 63 33) 48
34) 16 35) 100 36) 4
37) 110 38) 11 39) 42
40) 44 41) 3 42) 88
43) 40 44) 110 45) 14
46) 9 47) 20 48) 2
49) 50 50) 70 51) 9
52) 55 53) 6 54) 30
55) 120 56) 9 57) 144
58) 8 59) 110 60) 6

DAY 3
1) 100 2) 7 3) 9
4) 77 5) 18 6) 24
7) 36 8) 25 9) 35
10) 33 11) 9 12) 4
13) 16 14) 33 15) 32
16) 30 17) 32 18) 40
19) 24 20) 60 21) 9
22) 80 23) 110 24) 18
25) 1 26) 36 27) 42
28) 35 29) 36 30) 132
31) 81 32) 9 33) 72
34) 10 35) 121 36) 54
37) 6 38) 12 39) 27
40) 32 41) 132 42) 12
43) 16 44) 77 45) 70
46) 40 47) 21 48) 35
49) 40 50) 9 51) 45
52) 12 53) 36 54) 56
55) 18 56) 42 57) 132
58) 77 59) 24 60) 132

DAY 4
1) 54 2) 6 3) 8
4) 36 5) 110 6) 25
7) 40 8) 24 9) 16
10) 21 11) 27 12) 50
13) 24 14) 132 15) 56
16) 36 17) 54 18) 20
19) 8 20) 8 21) 5
22) 10 23) 40 24) 8
25) 80 26) 27 27) 5
28) 44 29) 4 30) 36
31) 11 32) 18 33) 20
34) 4 35) 48 36) 110
37) 55 38) 110 39) 10
40) 27 41) 25 42) 12
43) 60 44) 30 45) 8
46) 88 47) 42 48) 121
49) 2 50) 63 51) 88
52) 132 53) 50 54) 4
55) 56 56) 15 57) 56
58) 3 59) 48 60) 50

DAY 5
1) 18 2) 8 3) 42
4) 40 5) 110 6) 6
7) 12 8) 18 9) 8
10) 110 11) 63 12) 72
13) 4 14) 45 15) 54
16) 44 17) 108 18) 3
19) 24 20) 84 21) 56
22) 60 23) 5 24) 99
25) 12 26) 48 27) 7
28) 84 29) 10 30) 81
31) 56 32) 45 33) 120
34) 27 35) 5 36) 15
37) 28 38) 132 39) 12
40) 16 41) 8 42) 100
43) 40 44) 35 45) 22
46) 44 47) 25 48) 7
49) 48 50) 16 51) 60
52) 3 53) 30 54) 120
55) 30 56) 72 57) 77
58) 100 59) 30 60) 36

DAY 6
1) 22 2) 30 3) 63
4) 120 5) 100 6) 70
7) 6 8) 33 9) 9
10) 56 11) 24 12) 7
13) 20 14) 96 15) 96
16) 72 17) 54 18) 66
19) 25 20) 9 21) 11
22) 45 23) 144 24) 12
25) 21 26) 10 27) 24
28) 48 29) 6 30) 70
31) 108 32) 24 33) 33
34) 7 35) 4 36) 2
37) 84 38) 28 39) 40
40) 54 41) 121 42) 70
43) 7 44) 63 45) 24
46) 56 47) 120 48) 30
49) 11 50) 32 51) 42
52) 27 53) 4 54) 81
55) 60 56) 10 57) 40
58) 8 59) 32 60) 30

DAY 7
1) 72 2) 54 3) 72
4) 5 5) 77 6) 24
7) 120 8) 20 9) 18
10) 24 11) 14 12) 3
13) 40 14) 120 15) 40
16) 24 17) 8 18) 24
19) 36 20) 54 21) 60
22) 80 23) 24 24) 84
25) 44 26) 77 27) 10
28) 18 29) 20 30) 50
31) 10 32) 88 33) 14
34) 56 35) 14 36) 10
37) 20 38) 16 39) 121
40) 70 41) 108 42) 144
43) 18 44) 42 45) 12
46) 6 47) 5 48) 12
49) 24 50) 16 51) 64
52) 14 53) 96 54) 90
55) 16 56) 110 57) 40
58) 108 59) 88 60) 35

DAY 8
1) 56 2) 6 3) 30
4) 40 5) 6 6) 28
7) 5 8) 22 9) 50
10) 12 11) 6 12) 12
13) 30 14) 1 15) 8
16) 40 17) 11 18) 80
19) 90 20) 6 21) 45
22) 60 23) 70 24) 4
25) 45 26) 40 27) 144
28) 15 29) 12 30) 36
31) 54 32) 28 33) 32
34) 80 35) 28 36) 49
37) 11 38) 6 39) 96
40) 27 41) 6 42) 84
43) 108 44) 20 45) 84
46) 48 47) 20 48) 3
49) 110 50) 48 51) 8
52) 32 53) 6 54) 60
55) 56 56) 30 57) 56
58) 15 59) 16 60) 54

DAY 9
1) 45 2) 48 3) 12
4) 12 5) 24 6) 5
7) 18 8) 50 9) 63
10) 77 11) 100 12) 15
13) 48 14) 15 15) 36
16) 14 17) 30 18) 77
19) 18 20) 20 21) 88
22) 32 23) 45 24) 9
25) 55 26) 72 27) 36
28) 60 29) 2 30) 132
31) 12 32) 72 33) 36
34) 21 35) 30 36) 12
37) 120 38) 84 39) 110
40) 72 41) 10 42) 12
43) 72 44) 6 45) 48
46) 24 47) 11 48) 30
49) 4 50) 121 51) 77
52) 8 53) 33 54) 60
55) 40 56) 77 57) 28
58) 6 59) 77 60) 12

DAY 10
1) 12 2) 6 3) 66
4) 1 5) 108 6) 22
7) 7 8) 132 9) 14
10) 33 11) 8 12) 20
13) 64 14) 44 15) 60
16) 6 17) 36 18) 110
19) 32 20) 108 21) 77
22) 36 23) 9 24) 66
25) 8 26) 80 27) 22
28) 9 29) 10 30) 33
31) 63 32) 10 33) 40
34) 22 35) 63 36) 48
37) 54 38) 20 39) 99
40) 10 41) 30 42) 6
43) 45 44) 32 45) 55
46) 50 47) 5 48) 30
49) 3 50) 132 51) 9
52) 16 53) 6 54) 18
55) 6 56) 18 57) 45
58) 99 59) 15 60) 36

DAY 11
1) 72 2) 60 3) 49
4) 12 5) 120 6) 21
7) 60 8) 54 9) 60
10) 30 11) 21 12) 30
13) 10 14) 55 15) 24
16) 11 17) 16 18) 48
19) 120 20) 132 21) 9
22) 6 23) 35 24) 45
25) 2 26) 8 27) 48
28) 99 29) 8 30) 120
31) 8 32) 99 33) 27
34) 2 35) 99 36) 8
37) 32 38) 7 39) 8
40) 8 41) 60 42) 4
43) 63 44) 18 45) 6
46) 4 47) 77 48) 18
49) 8 50) 9 51) 48
52) 60 53) 4 54) 110
55) 40 56) 12 57) 32
58) 48 59) 18 60) 16

DAY 12
1) 63 2) 88 3) 10
4) 24 5) 21 6) 120
7) 99 8) 144 9) 42
10) 54 11) 15 12) 6
13) 60 14) 56 15) 28
16) 48 17) 55 18) 24
19) 48 20) 88 21) 84
22) 56 23) 120 24) 36
25) 121 26) 66 27) 20
28) 10 29) 90 30) 12
31) 6 32) 108 33) 72
34) 84 35) 132 36) 88
37) 144 38) 22 39) 18
40) 36 41) 40 42) 18
43) 24 44) 45 45) 7
46) 72 47) 77 48) 54
49) 6 50) 72 51) 16
52) 132 53) 84 54) 7
55) 10 56) 40 57) 32
58) 18 59) 5 60) 16

DAY 13
1) 16 2) 18 3) 40
4) 18 5) 48 6) 21
7) 24 8) 42 9) 14
10) 9 11) 55 12) 6
13) 24 14) 4 15) 88
16) 12 17) 36 18) 5
19) 24 20) 8 21) 24
22) 55 23) 10 24) 14
25) 48 26) 9 27) 32
28) 60 29) 16 30) 28
31) 36 32) 18 33) 44
34) 14 35) 24 36) 6
37) 5 38) 10 39) 54
40) 12 41) 12 42) 40
43) 20 44) 24 45) 55
46) 48 47) 60 48) 4
49) 24 50) 25 51) 40
52) 28 53) 21 54) 30
55) 72 56) 16 57) 20
58) 16 59) 15 60) 16

DAY 14
1) 14 2) 22 3) 88
4) 24 5) 96 6) 6
7) 49 8) 80 9) 28
10) 10 11) 84 12) 24
13) 72 14) 15 15) 90
16) 45 17) 84 18) 108
19) 60 20) 8 21) 55
22) 32 23) 12 24) 32
25) 20 26) 33 27) 108
28) 110 29) 36 30) 110
31) 12 32) 4 33) 96
34) 72 35) 9 36) 36
37) 42 38) 18 39) 16
40) 22 41) 88 42) 12
43) 70 44) 110 45) 60
46) 10 47) 28 48) 11
49) 63 50) 80 51) 12
52) 64 53) 70 54) 40
55) 63 56) 48 57) 96
58) 2 59) 33 60) 9

DAY 15
1) 80 2) 30 3) 7
4) 56 5) 72 6) 28
7) 84 8) 36 9) 33
10) 12 11) 70 12) 99
13) 21 14) 15 15) 121
16) 77 17) 99 18) 84
19) 45 20) 60 21) 18
22) 18 23) 66 24) 10
25) 121 26) 100 27) 33
28) 108 29) 50 30) 6
31) 4 32) 30 33) 72
34) 9 35) 30 36) 36
37) 3 38) 40 39) 77
40) 11 41) 66 42) 16
43) 90 44) 60 45) 36
46) 47 47) 88 48) 40
49) 8 50) 56 51) 88
52) 28 53) 40 54) 72
55) 108 56) 9 57) 21
58) 55 59) 27 60) 48

DAY 16
1) 45 2) 70 3) 77
4) 18 5) 4 6) 6
7) 72 8) 120 9) 77
10) 72 11) 5 12) 60
13) 21 14) 24 15) 30
16) 77 17) 25 18) 88
19) 64 20) 9 21) 56
22) 8 23) 28 24) 8
25) 21 26) 90 27) 20
28) 12 29) 66 30) 45
31) 4 32) 8 33) 55
34) 28 35) 25 36) 12
37) 42 38) 14 39) 42
40) 32 41) 40 42) 60
43) 120 44) 44 45) 77
46) 50 47) 96 48) 60
49) 60 50) 14 51) 54
52) 84 53) 21 54) 21
55) 36 56) 96 57) 18
58) 48 59) 80 60) 96

DAY 17
1) 20 2) 10 3) 36
4) 4 5) 12 6) 84
7) 10 8) 5 9) 12
10) 36 11) 12 12) 24
13) 24 14) 36 15) 18
16) 40 17) 36 18) 30
19) 12 20) 49 21) 33
22) 12 23) 11 24) 84
25) 22 26) 20 27) 110
28) 27 29) 60 30) 4
31) 44 32) 30 33) 3
34) 35 35) 27 36) 16
37) 66 38) 24 39) 40
40) 66 41) 54 42) 4
43) 80 44) 44 45) 63
46) 32 47) 12 48) 22
49) 22 50) 36 51) 20
52) 121 53) 7 54) 25
55) 11 56) 16 57) 15
58) 4 59) 120 60) 88

DAY 18
1) 84 2) 45 3) 27
4) 108 5) 50 6) 36
7) 63 8) 40 9) 42
10) 50 11) 84 12) 21
13) 45 14) 10 15) 108
16) 12 17) 70 18) 20
19) 36 20) 60 21) 12
22) 6 23) 3 24) 12
25) 8 26) 9 27) 90
28) 35 29) 9 30) 63
31) 70 32) 44 33) 66
34) 120 35) 66 36) 18
37) 18 38) 24 39) 42
40) 14 41) 80 42) 30
43) 10 44) 40 45) 36
46) 120 47) 10 48) 99
49) 6 50) 70 51) 81
52) 36 53) 50 54) 90
55) 56 56) 36 57) 3
58) 55 59) 24 60) 84

DAY 19
1) 72 2) 24 3) 11
4) 84 5) 1 6) 12
7) 63 8) 32 9) 22
10) 48 11) 60 12) 90
13) 121 14) 30 15) 20
16) 24 17) 24 18) 30
19) 48 20) 1 21) 42
22) 84 23) 40 24) 72
25) 63 26) 48 27) 20
28) 27 29) 40 30) 90
31) 80 32) 4 33) 72
34) 20 35) 32 36) 50
37) 48 38) 77 39) 50
40) 18 41) 24 42) 66
43) 81 44) 28 45) 12
46) 121 47) 18 48) 4
49) 40 50) 42 51) 3
52) 24 53) 90 54) 96
55) 36 56) 72 57) 24
58) 99 59) 121 60) 49

DAY 20
1) 80 2) 30 3) 7
4) 22 5) 24 6) 12
7) 96 8) 27 9) 64
10) 77 11) 6 12) 66
13) 9 14) 9 15) 24
16) 11 17) 16 18) 30
19) 24 20) 24 21) 99
22) 21 23) 5 24) 20
25) 60 26) 16 27) 108
28) 9 29) 14 30) 12
31) 72 32) 12 33) 60
34) 28 35) 96 36) 7
37) 18 38) 63 39) 80
40) 99 41) 81 42) 12
43) 55 44) 10 45) 33
46) 45 47) 88 48) 4
49) 44 50) 12 51) 24
52) 10 53) 7 54) 6
55) 66 56) 2 57) 9
58) 28 59) 18 60) 77

DAY 21
1) 60 2) 96 3) 44
4) 36 5) 32 6) 10
7) 12 8) 48 9) 60
10) 96 11) 80 12) 35
13) 70 14) 24 15) 12
16) 18 17) 48 18) 14
19) 33 20) 45 21) 55
22) 9 23) 7 24) 132
25) 18 26) 84 27) 22
28) 20 29) 45 30) 4
31) 63 32) 56 33) 12
34) 88 35) 80 36) 1
37) 88 38) 81 39) 22
40) 32 41) 63 42) 3
43) 88 44) 10 45) 24
46) 16 47) 28 48) 25
49) 12 50) 22 51) 10
52) 8 53) 12 54) 24
55) 18 56) 100 57) 15
58) 48 59) 22 60) 40

DAY 22
1) 120 2) 18 3) 15
4) 40 5) 18 6) 27
7) 2 8) 35 9) 21
10) 42 11) 30 12) 12
13) 72 14) 30 15) 4
16) 44 17) 45 18) 12
19) 30 20) 66 21) 30
22) 64 23) 14 24) 28
25) 72 26) 35 27) 4
28) 36 29) 36 30) 42
31) 8 32) 27 33) 70
34) 110 35) 60 36) 5
37) 20 38) 77 39) 20
40) 110 41) 8 42) 84
43) 81 44) 99 45) 24
46) 40 47) 96 48) 84
49) 7 50) 33 51) 16
52) 72 53) 45 54) 10
55) 72 56) 50 57) 4
58) 66 59) 12 60) 20

DAY 23
1) 64 2) 24 3) 5
4) 72 5) 8 6) 36
7) 1 8) 7 9) 24
10) 60 11) 8 12) 25
13) 70 14) 48 15) 28
16) 12 17) 24 18) 44
19) 56 20) 30 21) 45
22) 96 23) 49 24) 40
25) 63 26) 16 27) 3
28) 77 29) 70 30) 35
31) 1 32) 32 33) 36
34) 6 35) 3 36) 60
37) 60 38) 72 39) 48
40) 40 41) 80 42) 35
43) 21 44) 45 45) 80
46) 18 47) 36 48) 44
49) 60 50) 16 51) 10
52) 3 53) 84 54) 14
55) 40 56) 72 57) 32
58) 20 59) 27 60) 2

DAY 24
1) 90 2) 2 3) 1
4) 25 5) 45 6) 55
7) 24 8) 54 9) 48
10) 12 11) 56 12) 55
13) 80 14) 42 15) 11
16) 10 17) 48 18) 40
19) 55 20) 12 21) 44
22) 14 23) 48 24) 24
25) 72 26) 4 27) 96
28) 40 29) 44 30) 2
31) 24 32) 2 33) 30
34) 8 35) 9 36) 24
37) 72 38) 3 39) 5
40) 24 41) 54 42) 55
43) 14 44) 36 45) 84
46) 132 47) 1 48) 70
49) 18 50) 30 51) 120
52) 18 53) 15 54) 30
55) 3 56) 9 57) 24
58) 18 59) 4 60) 63

DAY 25
1) 60 2) 33 3) 8
4) 20 5) 30 6) 40
7) 12 8) 21 9) 48
10) 35 11) 54 12) 10
13) 10 14) 120 15) 30
16) 5 17) 10 18) 20
19) 30 20) 18 21) 18
22) 36 23) 1 24) 5
25) 11 26) 16 27) 27
28) 20 29) 35 30) 2
31) 48 32) 36 33) 10
34) 15 35) 16 36) 81
37) 8 38) 12 39) 33
40) 84 41) 108 42) 36
43) 36 44) 10 45) 72
46) 45 47) 8 48) 18
49) 44 50) 25 51) 90
52) 30 53) 4 54) 81
55) 9 56) 5 57) 33
58) 12 59) 5 60) 6

Multiplication Facts - Answer Key

DAY 26
1) 6 2) 40 3) 7 4) 3 5) 11 6) 4 7) 12 8) 40 9) 4 10) 24 11) 99 12) 8 13) 12 14) 42 15) 7 16) 120 17) 24 18) 10 19) 132 20) 18 21) 22 22) 96 23) 90 24) 3 25) 6 26) 48 27) 33 28) 50 29) 120 30) 66 31) 35 32) 50 33) 54 34) 12 35) 40 36) 56 37) 80 38) 48 39) 108 40) 50 41) 36 42) 10 43) 27 44) 15 45) 120 46) 30 47) 42 48) 36 49) 63 50) 12 51) 9 52) 8 53) 4 54) 55 55) 36 56) 5 57) 12 58) 16 59) 10 60) 40

DAY 27
1) 54 2) 49 3) 60 4) 48 5) 48 6) 10 7) 36 8) 9 9) 21 10) 21 11) 5 12) 66 13) 27 14) 24 15) 12 16) 120 17) 14 18) 12 19) 3 20) 121 21) 84 22) 72 23) 72 24) 10 25) 50 26) 12 27) 99 28) 48 29) 54 30) 3 31) 33 32) 132 33) 9 34) 8 35) 40 36) 40 37) 72 38) 14 39) 9 40) 90 41) 10 42) 18 43) 11 44) 35 45) 14 46) 55 47) 24 48) 6 49) 60 50) 15 51) 16 52) 48 53) 54 54) 48 55) 48 56) 22 57) 18 58) 63 59) 84 60) 27

DAY 28
1) 24 2) 11 3) 10 4) 27 5) 121 6) 12 7) 12 8) 36 9) 28 10) 5 11) 48 12) 60 13) 36 14) 10 15) 66 16) 12 17) 63 18) 20 19) 48 20) 11 21) 18 22) 63 23) 10 24) 8 25) 108 26) 14 27) 22 28) 56 29) 16 30) 6 31) 20 32) 33 33) 96 34) 77 35) 60 36) 14 37) 24 38) 12 39) 18 40) 40 41) 15 42) 66 43) 28 44) 1 45) 12 46) 60 47) 14 48) 30 49) 132 50) 35 51) 24 52) 5 53) 42 54) 72 55) 110 56) 18 57) 45 58) 27 59) 54 60) 42

DAY 29
1) 90 2) 99 3) 8 4) 30 5) 18 6) 25 7) 2 8) 44 9) 70 10) 36 11) 63 12) 60 13) 110 14) 15 15) 9 16) 60 17) 33 18) 10 19) 2 20) 4 21) 32 22) 110 23) 16 24) 56 25) 45 26) 33 27) 4 28) 33 29) 99 30) 108 31) 72 32) 40 33) 36 34) 108 35) 88 36) 28 37) 120 38) 80 39) 30 40) 63 41) 9 42) 5 43) 72 44) 49 45) 20 46) 96 47) 60 48) 88 49) 144 50) 36 51) 81 52) 8 53) 16 54) 24 55) 40 56) 60 57) 54 58) 6 59) 24 60) 60

DAY 30
1) 88 2) 18 3) 33 4) 100 5) 12 6) 16 7) 44 8) 50 9) 12 10) 7 11) 42 12) 7 13) 24 14) 1 15) 90 16) 12 17) 54 18) 35 19) 32 20) 12 21) 8 22) 63 23) 60 24) 4 25) 49 26) 9 27) 5 28) 99 29) 24 30) 2 31) 90 32) 132 33) 132 34) 6 35) 42 36) 132 37) 16 38) 36 39) 110 40) 48 41) 11 42) 4 43) 40 44) 9 45) 100 46) 10 47) 90 48) 96 49) 88 50) 77 51) 33 52) 5 53) 15 54) 80 55) 15 56) 72 57) 25 58) 11 59) 50 60) 8

DAY 31
1) 12 2) 121 3) 48 4) 144 5) 80 6) 22 7) 5 8) 55 9) 6 10) 108 11) 96 12) 22 13) 90 14) 77 15) 7 16) 80 17) 24 18) 33 19) 144 20) 132 21) 20 22) 55 23) 12 24) 55 25) 80 26) 70 27) 80 28) 99 29) 8 30) 33 31) 35 32) 21 33) 10 34) 108 35) 110 36) 144 37) 132 38) 144 39) 18 40) 54 41) 48 42) 6 43) 54 44) 11 45) 8 46) 49 47) 54 48) 3 49) 6 50) 15 51) 110 52) 40 53) 45 54) 63 55) 72 56) 33 57) 99 58) 21 59) 14 60) 30

DAY 32
1) 20 2) 18 3) 121 4) 9 5) 32 6) 4 7) 8 8) 10 9) 50 10) 14 11) 56 12) 12 13) 66 14) 99 15) 4 16) 100 17) 40 18) 4 19) 20 20) 4 21) 96 22) 3 23) 35 24) 60 25) 88 26) 24 27) 99 28) 24 29) 25 30) 18 31) 60 32) 28 33) 60 34) 27 35) 96 36) 30 37) 77 38) 30 39) 21 40) 56 41) 60 42) 120 43) 55 44) 45 45) 20 46) 25 47) 5 48) 30 49) 16 50) 3 51) 72 52) 15 53) 48 54) 32 55) 18 56) 6 57) 10 58) 21 59) 84 60) 42

DAY 33
1) 8 2) 99 3) 7 4) 80 5) 11 6) 25 7) 8 8) 15 9) 24 10) 12 11) 2 12) 49 13) 14 14) 54 15) 10 16) 72 17) 84 18) 110 19) 108 20) 28 21) 120 22) 11 23) 108 24) 35 25) 35 26) 48 27) 60 28) 64 29) 96 30) 32 31) 121 32) 36 33) 108 34) 99 35) 12 36) 40 37) 18 38) 16 39) 16 40) 55 41) 49 42) 66 43) 132 44) 100 45) 12 46) 12 47) 32 48) 6 49) 77 50) 60 51) 16 52) 15 53) 24 54) 45 55) 40 56) 90 57) 24 58) 40 59) 63 60) 72

DAY 34
1) 14 2) 8 3) 3 4) 108 5) 7 6) 48 7) 70 8) 24 9) 108 10) 36 11) 12 12) 88 13) 24 14) 99 15) 3 16) 4 17) 64 18) 80 19) 10 20) 2 21) 9 22) 11 23) 30 24) 60 25) 77 26) 12 27) 70 28) 7 29) 8 30) 63 31) 6 32) 22 33) 108 34) 66 35) 14 36) 48 37) 2 38) 70 39) 6 40) 6 41) 20 42) 6 43) 6 44) 27 45) 4 46) 32 47) 22 48) 12 49) 3 50) 2 51) 55 52) 12 53) 9 54) 8 55) 24 56) 80 57) 44 58) 44 59) 1 60) 8

DAY 35
1) 1 2) 15 3) 22 4) 144 5) 50 6) 63 7) 10 8) 40 9) 2 10) 40 11) 60 12) 28 13) 63 14) 24 15) 55 16) 80 17) 24 18) 90 19) 24 20) 70 21) 3 22) 8 23) 88 24) 99 25) 8 26) 48 27) 48 28) 35 29) 16 30) 60 31) 22 32) 12 33) 8 34) 7 35) 12 36) 36 37) 24 38) 77 39) 110 40) 8 41) 24 42) 80 43) 32 44) 8 45) 80 46) 27 47) 7 48) 8 49) 30 50) 5 51) 6 52) 7 53) 8 54) 40 55) 16 56) 18 57) 55 58) 60 59) 20 60) 30

DAY 36
1) 120 2) 40 3) 32 4) 108 5) 1 6) 108 7) 120 8) 10 9) 30 10) 9 11) 45 12) 18 13) 3 14) 10 15) 16 16) 24 17) 3 18) 27 19) 6 20) 81 21) 11 22) 44 23) 30 24) 7 25) 22 26) 99 27) 64 28) 84 29) 64 30) 20 31) 36 32) 8 33) 7 34) 16 35) 9 36) 2 37) 54 38) 15 39) 20 40) 63 41) 54 42) 28 43) 18 44) 72 45) 36 46) 50 47) 54 48) 24 49) 132 50) 50 51) 42 52) 12 53) 56 54) 20 55) 72 56) 35 57) 6 58) 22 59) 8 60) 20

DAY 37
1) 81 2) 2 3) 120 4) 30 5) 45 6) 18 7) 63 8) 24 9) 20 10) 8 11) 7 12) 99 13) 9 14) 33 15) 30 16) 27 17) 9 18) 36 19) 50 20) 6 21) 6 22) 72 23) 72 24) 9 25) 56 26) 21 27) 8 28) 5 29) 55 30) 27 31) 21 32) 24 33) 20 34) 20 35) 10 36) 60 37) 50 38) 20 39) 11 40) 18 41) 99 42) 12 43) 9 44) 12 45) 90 46) 8 47) 88 48) 24 49) 12 50) 64 51) 6 52) 110 53) 8 54) 11 55) 36 56) 70 57) 110 58) 63 59) 5 60) 36

DAY 38
1) 14 2) 22 3) 6 4) 20 5) 50 6) 27 7) 10 8) 54 9) 2 10) 49 11) 2 12) 54 13) 72 14) 88 15) 4 16) 110 17) 60 18) 18 19) 81 20) 45 21) 2 22) 12 23) 10 24) 10 25) 6 26) 90 27) 30 28) 36 29) 90 30) 120 31) 49 32) 80 33) 70 34) 2 35) 9 36) 49 37) 40 38) 36 39) 48 40) 66 41) 36 42) 11 43) 36 44) 35 45) 24 46) 6 47) 56 48) 42 49) 14 50) 22 51) 5 52) 55 53) 100 54) 36 55) 33 56) 21 57) 36 58) 45 59) 36 60) 36

DAY 39
1) 21 2) 72 3) 28 4) 77 5) 24 6) 12 7) 72 8) 120 9) 48 10) 66 11) 9 12) 44 13) 1 14) 27 15) 90 16) 54 17) 120 18) 20 19) 20 20) 27 21) 96 22) 108 23) 16 24) 48 25) 18 26) 50 27) 36 28) 44 29) 66 30) 14 31) 8 32) 36 33) 14 34) 12 35) 12 36) 96 37) 2 38) 21 39) 66 40) 40 41) 72 42) 18 43) 42 44) 28 45) 144 46) 20 47) 35 48) 100 49) 60 50) 50 51) 44 52) 108 53) 15 54) 120 55) 36 56) 36 57) 100 58) 88 59) 42 60) 25

DAY 40
1) 77 2) 36 3) 77 4) 45 5) 48 6) 11 7) 18 8) 36 9) 40 10) 24 11) 12 12) 108 13) 11 14) 60 15) 27 16) 40 17) 22 18) 4 19) 27 20) 15 21) 66 22) 12 23) 12 24) 36 25) 84 26) 3 27) 72 28) 99 29) 4 30) 20 31) 14 32) 90 33) 16 34) 9 35) 132 36) 16 37) 35 38) 88 39) 110 40) 24 41) 14 42) 8 43) 7 44) 11 45) 110 46) 77 47) 8 48) 36 49) 28 50) 32 51) 64 52) 144 53) 36 54) 36 55) 20 56) 14 57) 60 58) 18 59) 12 60) 66

DAY 41
1) 35 2) 80 3) 33 4) 90 5) 15 6) 70 7) 99 8) 84 9) 108 10) 18 11) 4 12) 24 13) 100 14) 3 15) 55 16) 18 17) 20 18) 70 19) 2 20) 99 21) 80 22) 9 23) 60 24) 110 25) 50 26) 44 27) 14 28) 24 29) 96 30) 1 31) 88 32) 9 33) 12 34) 30 35) 90 36) 42 37) 70 38) 12 39) 22 40) 40 41) 28 42) 42 43) 20 44) 18 45) 81 46) 56 47) 35 48) 24 49) 12 50) 20 51) 72 52) 45 53) 8 54) 60 55) 132 56) 100 57) 63 58) 40 59) 50 60) 81

DAY 42
1) 28 2) 55 3) 16 4) 16 5) 48 6) 60 7) 20 8) 110 9) 49 10) 21 11) 21 12) 96 13) 40 14) 32 15) 72 16) 132 17) 1 18) 48 19) 11 20) 16 21) 56 22) 50 23) 20 24) 144 25) 33 26) 35 27) 132 28) 48 29) 50 30) 80 31) 28 32) 32 33) 84 34) 55 35) 70 36) 48 37) 100 38) 49 39) 77 40) 12 41) 56 42) 48 43) 110 44) 27 45) 60 46) 40 47) 11 48) 12 49) 36 50) 80 51) 18 52) 2 53) 64 54) 96 55) 30 56) 24 57) 48 58) 88 59) 44 60) 12

DAY 43
1) 64 2) 90 3) 30 4) 7 5) 18 6) 32 7) 2 8) 81 9) 33 10) 42 11) 108 12) 90 13) 60 14) 1 15) 35 16) 30 17) 15 18) 6 19) 90 20) 96 21) 10 22) 48 23) 20 24) 132 25) 63 26) 55 27) 77 28) 4 29) 66 30) 27 31) 72 32) 50 33) 60 34) 35 35) 44 36) 9 37) 8 38) 88 39) 44 40) 84 41) 81 42) 20 43) 9 44) 70 45) 2 46) 10 47) 2 48) 28 49) 2 50) 21 51) 12 52) 40 53) 11 54) 72 55) 9 56) 77 57) 80 58) 8 59) 44 60) 12

DAY 44
1) 66 2) 21 3) 3 4) 33 5) 64 6) 81 7) 6 8) 40 9) 6 10) 44 11) 64 12) 45 13) 42 14) 56 15) 70 16) 32 17) 5 18) 90 19) 56 20) 8 21) 28 22) 24 23) 55 24) 44 25) 2 26) 72 27) 10 28) 12 29) 5 30) 72 31) 15 32) 33 33) 15 34) 88 35) 24 36) 8 37) 32 38) 40 39) 96 40) 55 41) 6 42) 1 43) 22 44) 11 45) 10 46) 36 47) 99 48) 9 49) 12 50) 12 51) 50 52) 4 53) 63 54) 8 55) 72 56) 20 57) 24 58) 63 59) 6 60) 4

DAY 45
1) 60 2) 108 3) 96 4) 56 5) 48 6) 9 7) 120 8) 24 9) 9 10) 2 11) 60 12) 20 13) 27 14) 88 15) 64 16) 70 17) 8 18) 48 19) 24 20) 36 21) 121 22) 10 23) 30 24) 2 25) 2 26) 14 27) 18 28) 20 29) 30 30) 99 31) 6 32) 45 33) 4 34) 9 35) 72 36) 12 37) 15 38) 90 39) 42 40) 36 41) 7 42) 4 43) 42 44) 132 45) 16 46) 77 47) 22 48) 9 49) 70 50) 132 51) 72 52) 6 53) 3 54) 80 55) 18 56) 8 57) 4 58) 10 59) 84 60) 64

DAY 46
1) 55 2) 20 3) 20 4) 100 5) 10 6) 44 7) 72 8) 27 9) 56 10) 132 11) 15 12) 96 13) 24 14) 24 15) 132 16) 96 17) 24 18) 132 19) 11 20) 24 21) 20 22) 7 23) 88 24) 32 25) 36 26) 48 27) 20 28) 30 29) 36 30) 16 31) 77 32) 33 33) 18 34) 88 35) 90 36) 63 37) 10 38) 36 39) 24 40) 49 41) 108 42) 11 43) 27 44) 8 45) 72 46) 35 47) 24 48) 18 49) 42 50) 21 51) 5 52) 48 53) 45 54) 18 55) 66 56) 20 57) 72 58) 70 59) 120 60) 8

DAY 47
1) 72 2) 36 3) 3 4) 11 5) 96 6) 25 7) 36 8) 88 9) 132 10) 20 11) 7 12) 36 13) 7 14) 90 15) 110 16) 30 17) 42 18) 48 19) 4 20) 66 21) 24 22) 72 23) 120 24) 4 25) 30 26) 22 27) 49 28) 20 29) 44 30) 30 31) 72 32) 36 33) 32 34) 22 35) 15 36) 80 37) 24 38) 28 39) 30 40) 30 41) 10 42) 28 43) 4 44) 28 45) 20 46) 32 47) 50 48) 20 49) 36 50) 18 51) 90 52) 2 53) 50 54) 56 55) 84 56) 40 57) 28 58) 132 59) 120 60) 14

DAY 48
1) 20 2) 22 3) 36 4) 77 5) 144 6) 36 7) 21 8) 64 9) 22 10) 32 11) 24 12) 10 13) 144 14) 56 15) 72 16) 40 17) 99 18) 72 19) 48 20) 88 21) 64 22) 60 23) 40 24) 42 25) 24 26) 18 27) 12 28) 12 29) 66 30) 30 31) 81 32) 48 33) 56 34) 30 35) 48 36) 12 37) 44 38) 21 39) 7 40) 120 41) 27 42) 90 43) 6 44) 110 45) 49 46) 40 47) 8 48) 35 49) 21 50) 55 51) 7 52) 10 53) 30 54) 16 55) 60 56) 35 57) 4 58) 60 59) 108 60) 9

DAY 49
1) 40 2) 70 3) 24 4) 70 5) 21 6) 6 7) 44 8) 21 9) 12 10) 36 11) 99 12) 35 13) 24 14) 8 15) 30 16) 96 17) 80 18) 12 19) 24 20) 90 21) 30 22) 12 23) 132 24) 48 25) 70 26) 36 27) 110 28) 20 29) 30 30) 60 31) 60 32) 4 33) 40 34) 54 35) 60 36) 6 37) 15 38) 56 39) 8 40) 18 41) 50 42) 108 43) 7 44) 90 45) 6 46) 66 47) 36 48) 24 49) 72 50) 60 51) 60 52) 5 53) 14 54) 12 55) 110 56) 40 57) 55 58) 56 59) 30 60) 66

DAY 50
1) 66 2) 120 3) 10 4) 88 5) 33 6) 6 7) 81 8) 80 9) 48 10) 30 11) 72 12) 6 13) 20 14) 100 15) 96 16) 25 17) 35 18) 24 19) 44 20) 35 21) 4 22) 6 23) 15 24) 63 25) 66 26) 10 27) 100 28) 144 29) 70 30) 36 31) 40 32) 21 33) 5 34) 56 35) 110 36) 10 37) 77 38) 99 39) 8 40) 11 41) 2 42) 132 43) 36 44) 40 45) 40 46) 36 47) 50 48) 48 49) 42 50) 33 51) 8 52) 7 53) 44 54) 70 55) 7 56) 90 57) 33 58) 72 59) 30 60) 56

Division Facts - Answer Key

DAY 51
1) 1 2) 2 3) 6
4) 1 5) 8 6) 8
7) 2 8) 9 9) 8
10) 12 11) 7 12) 4
13) 2 14) 3 15) 4
16) 4 17) 6 18) 8
19) 3 20) 3 21) 2
22) 7 23) 5 24) 6
25) 6 26) 7 27) 10
28) 3 29) 5 30) 1
31) 1 32) 10 33) 1
34) 11 35) 9 36) 1
37) 9 38) 9 39) 8
40) 12 41) 2 42) 9
43) 7 44) 8 45) 1
46) 4 47) 11 48) 9
49) 11 50) 12 51) 5
52) 4 53) 1 54) 4
55) 5 56) 10 57) 11
58) 1 59) 11 60) 7

DAY 52
1) 3 2) 6 3) 10
4) 6 5) 5 6) 7
7) 6 8) 6 9) 9
10) 4 11) 4 12) 3
13) 11 14) 5 15) 1
16) 11 17) 3 18) 4
19) 10 20) 7 21) 8
22) 7 23) 9 24) 2
25) 1 26) 6 27) 3
28) 6 29) 10 30) 2
31) 12 32) 12 33) 7
34) 7 35) 8 36) 12
37) 11 38) 12 39) 7
40) 10 41) 12 42) 11
43) 6 44) 6 45) 11
46) 5 47) 5 48) 1
49) 4 50) 11 51) 5
52) 6 53) 11 54) 3
55) 7 56) 9 57) 6
58) 5 59) 6 60) 1

DAY 53
1) 8 2) 6 3) 2
4) 10 5) 8 6) 8
7) 5 8) 1 9) 7
10) 10 11) 8 12) 9
13) 11 14) 11 15) 8
16) 4 17) 10 18) 9
19) 12 20) 6 21) 12
22) 8 23) 9 24) 7
25) 1 26) 3 27) 3
28) 1 29) 8 30) 10
31) 4 32) 7 33) 5
34) 4 35) 7 36) 7
37) 6 38) 6 39) 1
40) 7 41) 6 42) 2
43) 4 44) 8 45) 2
46) 12 47) 8 48) 9
49) 6 50) 12 51) 9
52) 2 53) 4 54) 12
55) 6 56) 4 57) 4
58) 10 59) 5 60) 4

DAY 54
1) 12 2) 3 3) 11
4) 1 5) 6 6) 11
7) 6 8) 5 9) 8
10) 9 11) 3 12) 12
13) 7 14) 2 15) 3
16) 4 17) 10 18) 12
19) 3 20) 2 21) 3
22) 4 23) 4 24) 10
25) 11 26) 8 27) 10
28) 10 29) 2 30) 8
31) 7 32) 1 33) 2
34) 7 35) 10 36) 2
37) 10 38) 10 39) 7
40) 5 41) 4 42) 9
43) 12 44) 3 45) 12
46) 5 47) 3 48) 5
49) 4 50) 11 51) 2
52) 5 53) 11 54) 2
55) 9 56) 1 57) 5
58) 12 59) 6 60) 8

DAY 55
1) 4 2) 11 3) 5
4) 8 5) 10 6) 12
7) 11 8) 1 9) 9
10) 6 11) 12 12) 2
13) 12 14) 7 15) 10
16) 6 17) 3 18) 7
19) 3 20) 4 21) 7
22) 12 23) 12 24) 9
25) 1 26) 8 27) 10
28) 10 29) 6 30) 1
31) 11 32) 3 33) 1
34) 12 35) 10 36) 5
37) 1 38) 1 39) 6
40) 12 41) 3 42) 2
43) 6 44) 8 45) 5
46) 5 47) 2 48) 10
49) 3 50) 7 51) 3
52) 7 53) 7 54) 5
55) 6 56) 1 57) 9
58) 8 59) 6 60) 1

DAY 56
1) 7 2) 4 3) 2
4) 4 5) 1 6) 5
7) 5 8) 10 9) 9
10) 7 11) 12 12) 5
13) 5 14) 2 15) 6
16) 7 17) 8 18) 7
19) 1 20) 8 21) 9
22) 2 23) 4 24) 3
25) 12 26) 2 27) 6
28) 11 29) 3 30) 11
31) 3 32) 10 33) 9
34) 12 35) 12 36) 11
37) 12 38) 5 39) 9
40) 12 41) 3 42) 8
43) 2 44) 2 45) 1
46) 1 47) 6 48) 4
49) 6 50) 12 51) 2
52) 6 53) 11 54) 2
55) 2 56) 1 57) 8
58) 9 59) 3 60) 6

DAY 57
1) 4 2) 10 3) 2
4) 4 5) 3 6) 9
7) 1 8) 4 9) 8
10) 10 11) 10 12) 6
13) 2 14) 6 15) 1
16) 11 17) 1 18) 2
19) 6 20) 5 21) 6
22) 1 23) 1 24) 5
25) 10 26) 10 27) 1
28) 7 29) 4 30) 11
31) 5 32) 12 33) 3
34) 3 35) 2 36) 2
37) 8 38) 8 39) 5
40) 1 41) 5 42) 1
43) 4 44) 8 45) 5
46) 4 47) 12 48) 8
49) 7 50) 7 51) 6
52) 10 53) 5 54) 2
55) 3 56) 5 57) 9
58) 9 59) 12 60) 8

DAY 58
1) 8 2) 8 3) 5
4) 12 5) 2 6) 8
7) 4 8) 10 9) 9
10) 1 11) 10 12) 8
13) 3 14) 5 15) 5
16) 3 17) 2 18) 4
19) 1 20) 8 21) 2
22) 1 23) 7 24) 7
25) 7 26) 12 27) 9
28) 11 29) 10 30) 5
31) 12 32) 8 33) 12
34) 5 35) 1 36) 2
37) 12 38) 9 39) 4
40) 7 41) 7 42) 7
43) 1 44) 12 45) 4
46) 8 47) 6 48) 10
49) 8 50) 12 51) 2
52) 10 53) 6 54) 11
55) 11 56) 7 57) 5
58) 11 59) 6 60) 4

DAY 59
1) 5 2) 6 3) 5
4) 8 5) 6 6) 4
7) 1 8) 9 9) 11
10) 1 11) 5 12) 1
13) 4 14) 2 15) 6
16) 2 17) 2 18) 5
19) 4 20) 7 21) 9
22) 12 23) 8 24) 3
25) 3 26) 10 27) 11
28) 3 29) 9 30) 10
31) 4 32) 1 33) 1
34) 9 35) 9 36) 5
37) 11 38) 9 39) 4
40) 6 41) 3 42) 3
43) 7 44) 4 45) 5
46) 12 47) 11 48) 9
49) 12 50) 2 51) 12
52) 4 53) 1 54) 2
55) 12 56) 1 57) 7
58) 9 59) 7 60) 3

DAY 60
1) 11 2) 3 3) 12
4) 10 5) 4 6) 1
7) 3 8) 11 9) 4
10) 3 11) 6 12) 7
13) 12 14) 1 15) 5
16) 6 17) 7 18) 5
19) 2 20) 6 21) 6
22) 9 23) 6 24) 1
25) 10 26) 7 27) 1
28) 5 29) 2 30) 7
31) 6 32) 7 33) 2
34) 7 35) 10 36) 11
37) 7 38) 8 39) 3
40) 2 41) 9 42) 4
43) 12 44) 11 45) 12
46) 10 47) 10 48) 4
49) 4 50) 12 51) 2
52) 1 53) 8 54) 2
55) 1 56) 11 57) 8
58) 7 59) 11 60) 12

DAY 61
1) 9 2) 8 3) 1
4) 9 5) 5 6) 7
7) 7 8) 4 9) 9
10) 10 11) 4 12) 1
13) 5 14) 4 15) 8
16) 12 17) 4 18) 6
19) 6 20) 11 21) 11
22) 6 23) 2 24) 3
25) 5 26) 1 27) 12
28) 6 29) 9 30) 2
31) 5 32) 8 33) 9
34) 1 35) 5 36) 10
37) 4 38) 1 39) 5
40) 4 41) 2 42) 11
43) 6 44) 7 45) 6
46) 8 47) 9 48) 7
49) 6 50) 2 51) 7
52) 11 53) 8 54) 11
55) 1 56) 8 57) 4
58) 2 59) 7 60) 10

DAY 62
1) 5 2) 1 3) 6
4) 10 5) 6 6) 8
7) 7 8) 11 9) 3
10) 3 11) 3 12) 9
13) 7 14) 11 15) 1
16) 7 17) 1 18) 12
19) 11 20) 1 21) 5
22) 4 23) 2 24) 2
25) 3 26) 2 27) 10
28) 9 29) 6 30) 10
31) 5 32) 8 33) 3
34) 6 35) 3 36) 9
37) 4 38) 1 39) 10
40) 5 41) 7 42) 1
43) 5 44) 3 45) 3
46) 12 47) 2 48) 2
49) 12 50) 7 51) 10
52) 12 53) 12 54) 11
55) 3 56) 11 57) 8
58) 2 59) 11 60) 3

DAY 63
1) 5 2) 6 3) 3
4) 10 5) 7 6) 9
7) 12 8) 6 9) 12
10) 2 11) 10 12) 4
13) 11 14) 1 15) 12
16) 12 17) 3 18) 10
19) 6 20) 9 21) 2
22) 12 23) 2 24) 4
25) 2 26) 6 27) 2
28) 9 29) 1 30) 5
31) 9 32) 1 33) 11
34) 10 35) 8 36) 2
37) 3 38) 10 39) 4
40) 10 41) 1 42) 2
43) 5 44) 2 45) 3
46) 12 47) 5 48) 8
49) 2 50) 8 51) 4
52) 1 53) 3 54) 10
55) 6 56) 1 57) 12
58) 8 59) 12 60) 2

DAY 64
1) 9 2) 6 3) 12
4) 2 5) 11 6) 7
7) 8 8) 9 9) 2
10) 8 11) 9 12) 4
13) 4 14) 12 15) 5
16) 4 17) 11 18) 10
19) 6 20) 3 21) 2
22) 9 23) 9 24) 1
25) 1 26) 9 27) 8
28) 9 29) 7 30) 11
31) 1 32) 4 33) 4
34) 10 35) 3 36) 6
37) 1 38) 8 39) 9
40) 9 41) 4 42) 7
43) 8 44) 11 45) 11
46) 3 47) 11 48) 5
49) 11 50) 12 51) 6
52) 11 53) 1 54) 3
55) 3 56) 11 57) 11
58) 8 59) 11 60) 1

DAY 65
1) 12 2) 7 3) 4
4) 6 5) 4 6) 11
7) 10 8) 4 9) 3
10) 5 11) 9 12) 1
13) 3 14) 6 15) 10
16) 4 17) 2 18) 4
19) 2 20) 6 21) 7
22) 10 23) 2 24) 7
25) 1 26) 1 27) 1
28) 3 29) 4 30) 9
31) 5 32) 6 33) 12
34) 6 35) 1 36) 10
37) 5 38) 6 39) 11
40) 3 41) 12 42) 12
43) 2 44) 2 45) 2
46) 10 47) 3 48) 4
49) 9 50) 8 51) 2
52) 11 53) 12 54) 1
55) 3 56) 5 57) 11
58) 1 59) 8 60) 7

DAY 66
1) 7 2) 4 3) 10
4) 10 5) 4 6) 9
7) 6 8) 10 9) 5
10) 3 11) 12 12) 7
13) 6 14) 11 15) 12
16) 5 17) 5 18) 5
19) 5 20) 5 21) 11
22) 7 23) 9 24) 3
25) 1 26) 12 27) 4
28) 1 29) 2 30) 8
31) 2 32) 5 33) 1
34) 4 35) 11 36) 11
37) 1 38) 8 39) 11
40) 3 41) 10 42) 5
43) 10 44) 8 45) 2
46) 1 47) 3 48) 7
49) 2 50) 3 51) 8
52) 4 53) 5 54) 7
55) 3 56) 8 57) 6
58) 10 59) 6 60) 11

DAY 67
1) 6 2) 8 3) 7
4) 5 5) 6 6) 8
7) 2 8) 7 9) 8
10) 4 11) 5 12) 4
13) 4 14) 10 15) 10
16) 4 17) 10 18) 8
19) 9 20) 8 21) 11
22) 7 23) 8 24) 12
25) 4 26) 3 27) 5
28) 2 29) 12 30) 6
31) 11 32) 6 33) 6
34) 2 35) 12 36) 6
37) 10 38) 6 39) 3
40) 10 41) 1 42) 12
43) 12 44) 1 45) 4
46) 11 47) 7 48) 6
49) 2 50) 2 51) 11
52) 9 53) 5 54) 5
55) 6 56) 4 57) 5
58) 3 59) 11 60) 3

DAY 68
1) 10 2) 2 3) 2
4) 8 5) 8 6) 3
7) 12 8) 6 9) 8
10) 8 11) 5 12) 5
13) 5 14) 10 15) 6
16) 9 17) 6 18) 1
19) 2 20) 11 21) 2
22) 1 23) 3 24) 4
25) 7 26) 12 27) 2
28) 4 29) 11 30) 2
31) 4 32) 11 33) 6
34) 4 35) 12 36) 2
37) 2 38) 11 39) 10
40) 6 41) 12 42) 11
43) 9 44) 11 45) 3
46) 2 47) 11 48) 9
49) 6 50) 12 51) 12
52) 9 53) 8 54) 5
55) 12 56) 4 57) 4
58) 8 59) 2 60) 3

DAY 69
1) 6 2) 6 3) 12
4) 8 5) 5 6) 1
7) 2 8) 3 9) 1
10) 7 11) 4 12) 7
13) 12 14) 11 15) 7
16) 11 17) 5 18) 7
19) 9 20) 2 21) 10
22) 3 23) 2 24) 5
25) 7 26) 3 27) 2
28) 6 29) 9 30) 2
31) 1 32) 11 33) 12
34) 11 35) 9 36) 1
37) 7 38) 10 39) 5
40) 2 41) 5 42) 11
43) 11 44) 2 45) 11
46) 8 47) 10 48) 8
49) 8 50) 1 51) 6
52) 3 53) 6 54) 9
55) 11 56) 3 57) 5
58) 7 59) 5 60) 8

DAY 70
1) 11 2) 7 3) 5
4) 6 5) 3 6) 11
7) 8 8) 7 9) 7
10) 11 11) 8 12) 8
13) 6 14) 5 15) 5
16) 9 17) 2 18) 4
19) 9 20) 2 21) 10
22) 7 23) 4 24) 10
25) 8 26) 11 27) 1
28) 5 29) 3 30) 9
31) 9 32) 1 33) 7
34) 8 35) 5 36) 9
37) 3 38) 7 39) 12
40) 1 41) 1 42) 1
43) 4 44) 4 45) 10
46) 1 47) 4 48) 4
49) 10 50) 6 51) 6
52) 8 53) 2 54) 5
55) 10 56) 6 57) 8
58) 7 59) 3 60) 5

DAY 71
1) 1 2) 10 3) 3
4) 3 5) 11 6) 12
7) 12 8) 2 9) 8
10) 7 11) 1 12) 9
13) 3 14) 7 15) 11
16) 4 17) 6 18) 7
19) 1 20) 3 21) 4
22) 1 23) 3 24) 1
25) 1 26) 4 27) 7
28) 11 29) 6 30) 3
31) 6 32) 1 33) 12
34) 7 35) 12 36) 3
37) 4 38) 5 39) 8
40) 4 41) 8 42) 5
43) 9 44) 10 45) 2
46) 3 47) 2 48) 11
49) 3 50) 2 51) 1
52) 5 53) 11 54) 12
55) 11 56) 4 57) 2
58) 3 59) 7 60) 6

DAY 72
1) 3 2) 11 3) 2
4) 4 5) 9 6) 12
7) 3 8) 10 9) 8
10) 8 11) 1 12) 7
13) 10 14) 8 15) 3
16) 1 17) 10 18) 9
19) 2 20) 5 21) 5
22) 12 23) 11 24) 6
25) 10 26) 11 27) 1
28) 9 29) 6 30) 1
31) 9 32) 11 33) 1
34) 8 35) 12 36) 5
37) 6 38) 4 39) 9
40) 11 41) 1 42) 1
43) 10 44) 9 45) 11
46) 8 47) 10 48) 6
49) 12 50) 9 51) 8
52) 3 53) 5 54) 11
55) 11 56) 10 57) 7
58) 2 59) 2 60) 12

DAY 73
1) 4 2) 9 3) 10
4) 11 5) 8 6) 2
7) 1 8) 8 9) 10
10) 2 11) 6 12) 3
13) 11 14) 12 15) 8
16) 7 17) 8 18) 2
19) 12 20) 3 21) 8
22) 4 23) 10 24) 2
25) 1 26) 6 27) 4
28) 11 29) 6 30) 2
31) 8 32) 5 33) 11
34) 2 35) 11 36) 7
37) 3 38) 6 39) 10
40) 4 41) 7 42) 5
43) 6 44) 4 45) 9
46) 5 47) 5 48) 3
49) 5 50) 8 51) 7
52) 9 53) 9 54) 4
55) 5 56) 12 57) 1
58) 12 59) 3 60) 1

DAY 74
1) 1 2) 7 3) 4
4) 1 5) 9 6) 8
7) 4 8) 6 9) 11
10) 3 11) 9 12) 4
13) 11 14) 9 15) 9
16) 12 17) 7 18) 10
19) 8 20) 6 21) 12
22) 3 23) 6 24) 9
25) 10 26) 9 27) 2
28) 6 29) 3 30) 9
31) 8 32) 2 33) 2
34) 5 35) 2 36) 12
37) 3 38) 5 39) 7
40) 1 41) 3 42) 1
43) 12 44) 10 45) 1
46) 8 47) 3 48) 7
49) 9 50) 6 51) 6
52) 3 53) 11 54) 7
55) 1 56) 6 57) 6
58) 2 59) 8 60) 1

DAY 75
1) 12 2) 6 3) 10
4) 9 5) 8 6) 8
7) 11 8) 4 9) 5
10) 6 11) 7 12) 6
13) 5 14) 9 15) 3
16) 3 17) 3 18) 4
19) 6 20) 4 21) 11
22) 10 23) 4 24) 7
25) 12 26) 12 27) 1
28) 1 29) 6 30) 3
31) 4 32) 12 33) 3
34) 12 35) 4 36) 9
37) 2 38) 10 39) 7
40) 4 41) 1 42) 1
43) 4 44) 6 45) 9
46) 3 47) 3 48) 7
49) 2 50) 3 51) 4
52) 1 53) 1 54) 5
55) 10 56) 6 57) 11
58) 11 59) 9 60) 6

Division Facts – Answer Key

DAY 76
1) 3, 2) 11, 3) 1, 4) 10, 5) 5, 6) 6, 7) 10, 8) 9, 9) 11, 10) 8, 11) 9, 12) 8, 13) 5, 14) 10, 15) 8, 16) 2, 17) 4, 18) 12, 19) 3, 20) 8, 21) 9, 22) 4, 23) 9, 24) 6, 25) 1, 26) 12, 27) 10, 28) 3, 29) 6, 30) 10, 31) 6, 32) 6, 33) 10, 34) 4, 35) 11, 36) 4, 37) 5, 38) 9, 39) 5, 40) 4, 41) 8, 42) 10, 43) 8, 44) 6, 45) 12, 46) 11, 47) 4, 48) 7, 49) 1, 50) 8, 51) 8, 52) 6, 53) 3, 54) 7, 55) 11, 56) 6, 57) 10, 58) 7, 59) 12, 60) 12

DAY 77
1) 4, 2) 11, 3) 1, 4) 7, 5) 10, 6) 10, 7) 12, 8) 10, 9) 8, 10) 5, 11) 1, 12) 3, 13) 3, 14) 7, 15) 4, 16) 3, 17) 9, 18) 8, 19) 12, 20) 6, 21) 4, 22) 6, 23) 9, 24) 8, 25) 10, 26) 2, 27) 11, 28) 12, 29) 11, 30) 8, 31) 3, 32) 4, 33) 11, 34) 11, 35) 11, 36) 9, 37) 4, 38) 4, 39) 2, 40) 7, 41) 11, 42) 4, 43) 9, 44) 7, 45) 11, 46) 1, 47) 10, 48) 9, 49) 12, 50) 9, 51) 5, 52) 1, 53) 3, 54) 6, 55) 7, 56) 6, 57) 7, 58) 2, 59) 2, 60) 5

DAY 78
1) 5, 2) 3, 3) 4, 4) 8, 5) 1, 6) 6, 7) 2, 8) 6, 9) 4, 10) 8, 11) 7, 12) 10, 13) 7, 14) 11, 15) 8, 16) 12, 17) 11, 18) 8, 19) 3, 20) 9, 21) 7, 22) 9, 23) 8, 24) 11, 25) 12, 26) 2, 27) 3, 28) 11, 29) 9, 30) 3, 31) 8, 32) 1, 33) 7, 34) 12, 35) 7, 36) 4, 37) 9, 38) 11, 39) 2, 40) 9, 41) 2, 42) 11, 43) 11, 44) 12, 45) 1, 46) 3, 47) 5, 48) 7, 49) 5, 50) 6, 51) 6, 52) 5, 53) 9, 54) 7, 55) 10, 56) 11, 57) 8, 58) 8, 59) 10, 60) 9

DAY 79
1) 12, 2) 7, 3) 4, 4) 6, 5) 10, 6) 11, 7) 9, 8) 1, 9) 7, 10) 9, 11) 11, 12) 11, 13) 1, 14) 9, 15) 4, 16) 4, 17) 7, 18) 9, 19) 8, 20) 5, 21) 1, 22) 9, 23) 3, 24) 12, 25) 11, 26) 3, 27) 10, 28) 10, 29) 11, 30) 1, 31) 10, 32) 10, 33) 4, 34) 2, 35) 9, 36) 9, 37) 5, 38) 5, 39) 1, 40) 3, 41) 5, 42) 8, 43) 11, 44) 12, 45) 8, 46) 8, 47) 4, 48) 7, 49) 9, 50) 9, 51) 6, 52) 10, 53) 11, 54) 1, 55) 11, 56) 7, 57) 6, 58) 7, 59) 11, 60) 8

DAY 80
1) 3, 2) 8, 3) 1, 4) 12, 5) 12, 6) 4, 7) 12, 8) 7, 9) 5, 10) 8, 11) 9, 12) 6, 13) 8, 14) 5, 15) 4, 16) 3, 17) 11, 18) 11, 19) 10, 20) 7, 21) 2, 22) 1, 23) 12, 24) 5, 25) 2, 26) 9, 27) 5, 28) 10, 29) 3, 30) 2, 31) 3, 32) 10, 33) 10, 34) 11, 35) 4, 36) 12, 37) 10, 38) 1, 39) 1, 40) 4, 41) 9, 42) 10, 43) 5, 44) 4, 45) 1, 46) 8, 47) 1, 48) 8, 49) 7, 50) 9, 51) 11, 52) 8, 53) 3, 54) 12, 55) 10, 56) 4, 57) 4, 58) 6, 59) 7, 60) 9

DAY 81
1) 2, 2) 8, 3) 7, 4) 5, 5) 6, 6) 8, 7) 7, 8) 1, 9) 3, 10) 10, 11) 4, 12) 1, 13) 2, 14) 12, 15) 3, 16) 5, 17) 1, 18) 5, 19) 11, 20) 2, 21) 7, 22) 4, 23) 6, 24) 2, 25) 12, 26) 1, 27) 10, 28) 6, 29) 8, 30) 4, 31) 7, 32) 9, 33) 7, 34) 1, 35) 2, 36) 8, 37) 7, 38) 7, 39) 7, 40) 4, 41) 12, 42) 4, 43) 7, 44) 12, 45) 1, 46) 6, 47) 8, 48) 4, 49) 8, 50) 11, 51) 6, 52) 12, 53) 9, 54) 12, 55) 8, 56) 5, 57) 10, 58) 12, 59) 7, 60) 7

DAY 82
1) 12, 2) 7, 3) 11, 4) 4, 5) 1, 6) 9, 7) 2, 8) 1, 9) 5, 10) 3, 11) 7, 12) 4, 13) 11, 14) 2, 15) 2, 16) 7, 17) 8, 18) 11, 19) 9, 20) 12, 21) 3, 22) 8, 23) 4, 24) 11, 25) 4, 26) 1, 27) 12, 28) 2, 29) 10, 30) 2, 31) 5, 32) 10, 33) 5, 34) 1, 35) 5, 36) 4, 37) 4, 38) 11, 39) 8, 40) 3, 41) 8, 42) 5, 43) 8, 44) 9, 45) 12, 46) 9, 47) 1, 48) 1, 49) 4, 50) 4, 51) 2, 52) 12, 53) 10, 54) 2, 55) 8, 56) 4, 57) 2, 58) 8, 59) 11, 60) 2

DAY 83
1) 9, 2) 10, 3) 6, 4) 3, 5) 9, 6) 4, 7) 4, 8) 9, 9) 9, 10) 6, 11) 7, 12) 8, 13) 3, 14) 4, 15) 11, 16) 12, 17) 12, 18) 2, 19) 12, 20) 7, 21) 11, 22) 10, 23) 1, 24) 7, 25) 5, 26) 9, 27) 10, 28) 11, 29) 7, 30) 7, 31) 12, 32) 11, 33) 2, 34) 6, 35) 3, 36) 9, 37) 4, 38) 8, 39) 2, 40) 8, 41) 7, 42) 10, 43) 1, 44) 4, 45) 6, 46) 10, 47) 3, 48) 9, 49) 7, 50) 9, 51) 2, 52) 7, 53) 5, 54) 10, 55) 7, 56) 12, 57) 1, 58) 4, 59) 11, 60) 9

DAY 84
1) 9, 2) 9, 3) 4, 4) 12, 5) 7, 6) 3, 7) 6, 8) 1, 9) 10, 10) 5, 11) 3, 12) 9, 13) 2, 14) 12, 15) 11, 16) 8, 17) 4, 18) 3, 19) 8, 20) 4, 21) 3, 22) 10, 23) 1, 24) 6, 25) 11, 26) 5, 27) 9, 28) 9, 29) 2, 30) 11, 31) 5, 32) 10, 33) 1, 34) 7, 35) 4, 36) 7, 37) 1, 38) 2, 39) 6, 40) 3, 41) 8, 42) 12, 43) 3, 44) 3, 45) 9, 46) 12, 47) 3, 48) 10, 49) 5, 50) 4, 51) 9, 52) 2, 53) 9, 54) 7, 55) 4, 56) 6, 57) 11, 58) 7, 59) 6, 60) 7

DAY 85
1) 7, 2) 6, 3) 7, 4) 6, 5) 8, 6) 9, 7) 12, 8) 7, 9) 5, 10) 9, 11) 8, 12) 5, 13) 8, 14) 5, 15) 7, 16) 8, 17) 2, 18) 5, 19) 1, 20) 5, 21) 6, 22) 6, 23) 11, 24) 3, 25) 1, 26) 1, 27) 6, 28) 6, 29) 12, 30) 11, 31) 8, 32) 2, 33) 11, 34) 4, 35) 6, 36) 4, 37) 5, 38) 6, 39) 1, 40) 1, 41) 8, 42) 7, 43) 12, 44) 4, 45) 5, 46) 7, 47) 7, 48) 7, 49) 10, 50) 7, 51) 10, 52) 7, 53) 7, 54) 7, 55) 5, 56) 9, 57) 4, 58) 4, 59) 3, 60) 7

DAY 86
1) 8, 2) 3, 3) 11, 4) 3, 5) 3, 6) 1, 7) 9, 8) 6, 9) 8, 10) 3, 11) 6, 12) 12, 13) 6, 14) 6, 15) 5, 16) 7, 17) 12, 18) 7, 19) 2, 20) 8, 21) 11, 22) 2, 23) 5, 24) 6, 25) 4, 26) 4, 27) 8, 28) 8, 29) 11, 30) 3, 31) 1, 32) 9, 33) 6, 34) 5, 35) 11, 36) 10, 37) 8, 38) 6, 39) 3, 40) 6, 41) 10, 42) 12, 43) 2, 44) 8, 45) 1, 46) 12, 47) 5, 48) 12, 49) 6, 50) 6, 51) 10, 52) 1, 53) 3, 54) 5, 55) 2, 56) 4, 57) 7, 58) 2, 59) 4, 60) 12

DAY 87
1) 5, 2) 2, 3) 3, 4) 5, 5) 1, 6) 11, 7) 7, 8) 12, 9) 11, 10) 7, 11) 10, 12) 9, 13) 1, 14) 2, 15) 8, 16) 7, 17) 3, 18) 6, 19) 10, 20) 10, 21) 4, 22) 11, 23) 6, 24) 8, 25) 2, 26) 7, 27) 1, 28) 8, 29) 1, 30) 6, 31) 7, 32) 8, 33) 10, 34) 3, 35) 11, 36) 11, 37) 8, 38) 11, 39) 10, 40) 4, 41) 11, 42) 2, 43) 3, 44) 7, 45) 7, 46) 4, 47) 8, 48) 1, 49) 1, 50) 11, 51) 7, 52) 6, 53) 2, 54) 2, 55) 10, 56) 8, 57) 5, 58) 2, 59) 4, 60) 12

DAY 88
1) 9, 2) 3, 3) 8, 4) 11, 5) 11, 6) 8, 7) 2, 8) 10, 9) 2, 10) 1, 11) 5, 12) 5, 13) 6, 14) 9, 15) 5, 16) 2, 17) 10, 18) 4, 19) 3, 20) 9, 21) 3, 22) 10, 23) 8, 24) 1, 25) 5, 26) 1, 27) 1, 28) 5, 29) 5, 30) 2, 31) 8, 32) 1, 33) 10, 34) 7, 35) 7, 36) 5, 37) 12, 38) 9, 39) 1, 40) 5, 41) 11, 42) 12, 43) 10, 44) 4, 45) 7, 46) 5, 47) 2, 48) 9, 49) 5, 50) 10, 51) 9, 52) 12, 53) 12, 54) 1, 55) 4, 56) 3, 57) 6, 58) 12, 59) 6, 60) 6

DAY 89
1) 2, 2) 4, 3) 8, 4) 10, 5) 11, 6) 2, 7) 1, 8) 4, 9) 12, 10) 2, 11) 11, 12) 12, 13) 9, 14) 6, 15) 2, 16) 8, 17) 10, 18) 5, 19) 9, 20) 10, 21) 7, 22) 4, 23) 3, 24) 4, 25) 2, 26) 12, 27) 11, 28) 8, 29) 9, 30) 10, 31) 4, 32) 6, 33) 2, 34) 7, 35) 2, 36) 8, 37) 7, 38) 8, 39) 8, 40) 5, 41) 4, 42) 7, 43) 9, 44) 11, 45) 9, 46) 12, 47) 6, 48) 5, 49) 5, 50) 10, 51) 7, 52) 12, 53) 3, 54) 12, 55) 5, 56) 6, 57) 4, 58) 5, 59) 6, 60) 5

DAY 90
1) 7, 2) 6, 3) 8, 4) 5, 5) 10, 6) 5, 7) 12, 8) 5, 9) 9, 10) 4, 11) 12, 12) 11, 13) 12, 14) 4, 15) 11, 16) 9, 17) 3, 18) 5, 19) 1, 20) 7, 21) 9, 22) 6, 23) 10, 24) 1, 25) 6, 26) 11, 27) 6, 28) 2, 29) 9, 30) 11, 31) 1, 32) 8, 33) 1, 34) 9, 35) 3, 36) 6, 37) 9, 38) 10, 39) 1, 40) 2, 41) 2, 42) 7, 43) 10, 44) 1, 45) 8, 46) 2, 47) 9, 48) 5, 49) 5, 50) 7, 51) 3, 52) 12, 53) 12, 54) 4, 55) 5, 56) 8, 57) 11, 58) 1, 59) 2, 60) 7

DAY 91
1) 6, 2) 8, 3) 6, 4) 12, 5) 5, 6) 10, 7) 12, 8) 7, 9) 8, 10) 2, 11) 6, 12) 5, 13) 6, 14) 4, 15) 10, 16) 1, 17) 3, 18) 7, 19) 5, 20) 9, 21) 7, 22) 3, 23) 8, 24) 1, 25) 5, 26) 10, 27) 4, 28) 12, 29) 10, 30) 6, 31) 3, 32) 6, 33) 9, 34) 4, 35) 11, 36) 1, 37) 8, 38) 10, 39) 11, 40) 1, 41) 1, 42) 9, 43) 3, 44) 1, 45) 3, 46) 3, 47) 6, 48) 1, 49) 2, 50) 6, 51) 8, 52) 6, 53) 3, 54) 4, 55) 12, 56) 3, 57) 7, 58) 8, 59) 10, 60) 1

DAY 92
1) 1, 2) 1, 3) 10, 4) 1, 5) 9, 6) 10, 7) 8, 8) 4, 9) 8, 10) 12, 11) 10, 12) 5, 13) 11, 14) 7, 15) 10, 16) 4, 17) 3, 18) 2, 19) 8, 20) 3, 21) 8, 22) 8, 23) 11, 24) 8, 25) 5, 26) 4, 27) 1, 28) 6, 29) 1, 30) 9, 31) 2, 32) 8, 33) 12, 34) 6, 35) 7, 36) 5, 37) 10, 38) 2, 39) 3, 40) 2, 41) 4, 42) 5, 43) 5, 44) 4, 45) 5, 46) 5, 47) 7, 48) 8, 49) 6, 50) 3, 51) 10, 52) 2, 53) 4, 54) 5, 55) 11, 56) 2, 57) 5, 58) 12, 59) 11, 60) 6

DAY 93
1) 5, 2) 12, 3) 4, 4) 1, 5) 6, 6) 8, 7) 3, 8) 8, 9) 4, 10) 4, 11) 9, 12) 2, 13) 7, 14) 6, 15) 4, 16) 4, 17) 3, 18) 4, 19) 11, 20) 12, 21) 2, 22) 6, 23) 8, 24) 3, 25) 3, 26) 1, 27) 12, 28) 5, 29) 10, 30) 10, 31) 12, 32) 9, 33) 8, 34) 11, 35) 9, 36) 6, 37) 1, 38) 11, 39) 4, 40) 9, 41) 1, 42) 10, 43) 1, 44) 1, 45) 1, 46) 2, 47) 6, 48) 9, 49) 5, 50) 8, 51) 9, 52) 7, 53) 4, 54) 6, 55) 7, 56) 5, 57) 7, 58) 2, 59) 4, 60) 6

DAY 94
1) 8, 2) 5, 3) 7, 4) 11, 5) 1, 6) 4, 7) 12, 8) 2, 9) 9, 10) 10, 11) 9, 12) 12, 13) 2, 14) 5, 15) 8, 16) 10, 17) 7, 18) 8, 19) 2, 20) 5, 21) 5, 22) 8, 23) 6, 24) 11, 25) 5, 26) 10, 27) 8, 28) 1, 29) 7, 30) 10, 31) 11, 32) 8, 33) 2, 34) 9, 35) 7, 36) 9, 37) 1, 38) 3, 39) 9, 40) 8, 41) 2, 42) 6, 43) 10, 44) 4, 45) 9, 46) 10, 47) 11, 48) 8, 49) 9, 50) 2, 51) 8, 52) 8, 53) 12, 54) 3, 55) 5, 56) 2, 57) 2, 58) 12, 59) 7, 60) 6

DAY 95
1) 5, 2) 8, 3) 11, 4) 5, 5) 9, 6) 5, 7) 9, 8) 2, 9) 3, 10) 10, 11) 12, 12) 6, 13) 2, 14) 7, 15) 5, 16) 12, 17) 10, 18) 2, 19) 2, 20) 6, 21) 7, 22) 6, 23) 5, 24) 3, 25) 6, 26) 10, 27) 2, 28) 11, 29) 9, 30) 9, 31) 10, 32) 1, 33) 6, 34) 12, 35) 4, 36) 7, 37) 6, 38) 7, 39) 7, 40) 5, 41) 12, 42) 7, 43) 6, 44) 11, 45) 8, 46) 2, 47) 7, 48) 7, 49) 2, 50) 1, 51) 10, 52) 5, 53) 2, 54) 12, 55) 8, 56) 9, 57) 7, 58) 10, 59) 7, 60) 7

DAY 96
1) 10, 2) 11, 3) 12, 4) 7, 5) 12, 6) 3, 7) 7, 8) 5, 9) 5, 10) 7, 11) 1, 12) 10, 13) 12, 14) 11, 15) 3, 16) 5, 17) 1, 18) 2, 19) 7, 20) 1, 21) 5, 22) 10, 23) 3, 24) 4, 25) 5, 26) 8, 27) 10, 28) 3, 29) 12, 30) 3, 31) 2, 32) 9, 33) 2, 34) 3, 35) 10, 36) 2, 37) 3, 38) 10, 39) 7, 40) 9, 41) 3, 42) 3, 43) 8, 44) 7, 45) 9, 46) 3, 47) 7, 48) 3, 49) 9, 50) 12, 51) 9, 52) 8, 53) 4, 54) 3, 55) 2, 56) 4, 57) 12, 58) 10, 59) 2, 60) 10

DAY 97
1) 8, 2) 5, 3) 3, 4) 10, 5) 10, 6) 1, 7) 5, 8) 8, 9) 7, 10) 7, 11) 5, 12) 4, 13) 6, 14) 3, 15) 7, 16) 3, 17) 5, 18) 4, 19) 6, 20) 3, 21) 4, 22) 7, 23) 2, 24) 12, 25) 7, 26) 12, 27) 5, 28) 11, 29) 3, 30) 9, 31) 7, 32) 11, 33) 9, 34) 10, 35) 11, 36) 10, 37) 12, 38) 10, 39) 6, 40) 11, 41) 4, 42) 8, 43) 12, 44) 7, 45) 4, 46) 9, 47) 11, 48) 3, 49) 1, 50) 5, 51) 6, 52) 12, 53) 7, 54) 12, 55) 2, 56) 2, 57) 1, 58) 1, 59) 9, 60) 7

DAY 98
1) 9, 2) 11, 3) 1, 4) 4, 5) 10, 6) 3, 7) 8, 8) 8, 9) 3, 10) 11, 11) 4, 12) 9, 13) 7, 14) 11, 15) 2, 16) 10, 17) 2, 18) 5, 19) 10, 20) 3, 21) 6, 22) 1, 23) 9, 24) 8, 25) 1, 26) 6, 27) 4, 28) 7, 29) 7, 30) 3, 31) 8, 32) 11, 33) 11, 34) 11, 35) 12, 36) 7, 37) 5, 38) 1, 39) 10, 40) 5, 41) 2, 42) 2, 43) 11, 44) 9, 45) 2, 46) 9, 47) 9, 48) 7, 49) 4, 50) 3, 51) 9, 52) 5, 53) 8, 54) 6, 55) 6, 56) 6, 57) 3, 58) 5, 59) 7, 60) 11

DAY 99
1) 10, 2) 12, 3) 7, 4) 3, 5) 10, 6) 10, 7) 2, 8) 7, 9) 12, 10) 9, 11) 1, 12) 6, 13) 11, 14) 7, 15) 12, 16) 10, 17) 3, 18) 7, 19) 3, 20) 2, 21) 6, 22) 1, 23) 1, 24) 5, 25) 8, 26) 10, 27) 7, 28) 8, 29) 10, 30) 7, 31) 8, 32) 4, 33) 1, 34) 6, 35) 5, 36) 2, 37) 2, 38) 2, 39) 3, 40) 4, 41) 10, 42) 8, 43) 9, 44) 1, 45) 6, 46) 8, 47) 12, 48) 9, 49) 4, 50) 11, 51) 6, 52) 12, 53) 6, 54) 1, 55) 12, 56) 6, 57) 10, 58) 10, 59) 2, 60) 6

DAY 100
1) 11, 2) 2, 3) 12, 4) 2, 5) 6, 6) 1, 7) 4, 8) 10, 9) 12, 10) 1, 11) 9, 12) 10, 13) 5, 14) 5, 15) 6, 16) 9, 17) 10, 18) 12, 19) 10, 20) 8, 21) 8, 22) 7, 23) 10, 24) 7, 25) 5, 26) 9, 27) 7, 28) 4, 29) 6, 30) 3, 31) 2, 32) 10, 33) 12, 34) 11, 35) 8, 36) 2, 37) 8, 38) 3, 39) 4, 40) 12, 41) 10, 42) 3, 43) 12, 44) 2, 45) 1, 46) 7, 47) 12, 48) 9, 49) 3, 50) 4, 51) 12, 52) 10, 53) 4, 54) 4, 55) 2, 56) 1, 57) 12, 58) 6, 59) 1, 60) 3

Multiplication Tables

1 Times Table
1 x 1 = 1
1 x 2 = 2
1 x 3 = 3
1 x 4 = 4
1 x 5 = 5
1 x 6 = 6
1 x 7 = 7
1 x 8 = 8
1 x 9 = 9
1 x 10 = 10
1 x 11 = 11
1 x 12 = 12

2 Times Table
2 x 1 = 2
2 x 2 = 4
2 x 3 = 6
2 x 4 = 8
2 x 5 = 10
2 x 6 = 12
2 x 7 = 14
2 x 8 = 16
2 x 9 = 18
2 x 10 = 20
2 x 11 = 22
2 x 12 = 24

3 Times Table
3 x 1 = 3
3 x 2 = 6
3 x 3 = 9
3 x 4 = 12
3 x 5 = 15
3 x 6 = 18
3 x 7 = 21
3 x 8 = 24
3 x 9 = 27
3 x 10 = 30
3 x 11 = 33
3 x 12 = 36

4 Times Table
4 x 1 = 4
4 x 2 = 8
4 x 3 = 12
4 x 4 = 16
4 x 5 = 20
4 x 6 = 24
4 x 7 = 28
4 x 8 = 32
4 x 9 = 36
4 x 10 = 40
4 x 11 = 44
4 x 12 = 48

5 Times Table
5 x 1 = 5
5 x 2 = 10
5 x 3 = 15
5 x 4 = 20
5 x 5 = 25
5 x 6 = 30
5 x 7 = 35
5 x 8 = 40
5 x 9 = 45
5 x 10 = 50
5 x 11 = 55
5 x 12 = 60

6 Times Table
6 x 1 = 6
6 x 2 = 12
6 x 3 = 18
6 x 4 = 24
6 x 5 = 30
6 x 6 = 36
6 x 7 = 42
6 x 8 = 48
6 x 9 = 54
6 x 10 = 60
6 x 11 = 66
6 x 12 = 72

7 Times Table
7 x 1 = 7
7 x 2 = 14
7 x 3 = 21
7 x 4 = 28
7 x 5 = 35
7 x 6 = 42
7 x 7 = 49
7 x 8 = 56
7 x 9 = 63
7 x 10 = 70
7 x 11 = 77
7 x 12 = 84

8 Times Table
8 x 1 = 8
8 x 2 = 16
8 x 3 = 24
8 x 4 = 32
8 x 5 = 40
8 x 6 = 48
8 x 7 = 56
8 x 8 = 64
8 x 9 = 72
8 x 10 = 80
8 x 11 = 88
8 x 12 = 96

9 Times Table
9 x 1 = 9
9 x 2 = 18
9 x 3 = 27
9 x 4 = 36
9 x 5 = 45
9 x 6 = 54
9 x 7 = 63
9 x 8 = 72
9 x 9 = 81
9 x 10 = 90
9 x 11 = 99
9 x 12 = 108

10 Times Table
10 x 1 = 10
10 x 2 = 20
10 x 3 = 30
10 x 4 = 40
10 x 5 = 50
10 x 6 = 60
10 x 7 = 70
10 x 8 = 80
10 x 9 = 90
10 x 10 = 100
10 x 11 = 110
10 x 12 = 120

11 Times Table
11 x 1 = 11
11 x 2 = 22
11 x 3 = 33
11 x 4 = 44
11 x 5 = 55
11 x 6 = 66
11 x 7 = 77
11 x 8 = 88
11 x 9 = 99
11 x 10 = 110
11 x 11 = 121
11 x 12 = 132

12 Times Table
12 x 1 = 12
12 x 2 = 24
12 x 3 = 36
12 x 4 = 48
12 x 5 = 60
12 x 6 = 72
12 x 7 = 84
12 x 8 = 96
12 x 9 = 108
12 x 10 = 120
12 x 11 = 132
12 x 12 = 144

Division Tables

Dividing by 1
1 ÷ 1 = 1
2 ÷ 1 = 2
3 ÷ 1 = 3
4 ÷ 1 = 4
5 ÷ 1 = 5
6 ÷ 1 = 6
7 ÷ 1 = 7
8 ÷ 1 = 8
9 ÷ 1 = 9
10 ÷ 1 = 10
11 ÷ 1 = 11
12 ÷ 1 = 12

Dividing by 2
2 ÷ 2 = 1
4 ÷ 2 = 2
6 ÷ 2 = 3
8 ÷ 2 = 4
10 ÷ 2 = 5
12 ÷ 2 = 6
14 ÷ 2 = 7
16 ÷ 2 = 8
18 ÷ 2 = 9
20 ÷ 2 = 10
22 ÷ 2 = 11
24 ÷ 2 = 12

Dividing by 3
3 ÷ 3 = 1
6 ÷ 3 = 2
9 ÷ 3 = 3
12 ÷ 3 = 4
15 ÷ 3 = 5
18 ÷ 3 = 6
21 ÷ 3 = 7
24 ÷ 3 = 8
27 ÷ 3 = 9
30 ÷ 3 = 10
33 ÷ 3 = 11
36 ÷ 3 = 12

Dividing by 4
4 ÷ 4 = 1
8 ÷ 4 = 2
12 ÷ 4 = 3
16 ÷ 4 = 4
20 ÷ 4 = 5
24 ÷ 4 = 6
28 ÷ 4 = 7
32 ÷ 4 = 8
36 ÷ 4 = 9
40 ÷ 4 = 10
44 ÷ 4 = 11
48 ÷ 4 = 12

Dividing by 5
5 ÷ 5 = 1
10 ÷ 5 = 2
15 ÷ 5 = 3
20 ÷ 5 = 4
25 ÷ 5 = 5
30 ÷ 5 = 6
35 ÷ 5 = 7
40 ÷ 5 = 8
45 ÷ 5 = 9
50 ÷ 5 = 10
55 ÷ 5 = 11
60 ÷ 5 = 12

Dividing by 6
6 ÷ 6 = 1
12 ÷ 6 = 2
18 ÷ 6 = 3
24 ÷ 6 = 4
30 ÷ 6 = 5
36 ÷ 6 = 6
42 ÷ 6 = 7
48 ÷ 6 = 8
54 ÷ 6 = 9
60 ÷ 6 = 10
66 ÷ 6 = 11
72 ÷ 6 = 12

Dividing by 7
7 ÷ 7 = 1
14 ÷ 7 = 2
21 ÷ 7 = 3
28 ÷ 7 = 4
35 ÷ 7 = 5
42 ÷ 7 = 6
49 ÷ 7 = 7
56 ÷ 7 = 8
63 ÷ 7 = 9
70 ÷ 7 = 10
77 ÷ 7 = 11
84 ÷ 7 = 12

Dividing by 8
8 ÷ 8 = 1
16 ÷ 8 = 2
24 ÷ 8 = 3
32 ÷ 8 = 4
40 ÷ 8 = 5
48 ÷ 8 = 6
56 ÷ 8 = 7
64 ÷ 8 = 8
72 ÷ 8 = 9
80 ÷ 8 = 10
88 ÷ 8 = 11
96 ÷ 8 = 12

Dividing by 9
9 ÷ 9 = 1
18 ÷ 9 = 2
27 ÷ 9 = 3
36 ÷ 9 = 4
45 ÷ 9 = 5
54 ÷ 9 = 6
63 ÷ 9 = 7
72 ÷ 9 = 8
81 ÷ 9 = 9
90 ÷ 9 = 10
99 ÷ 9 = 11
108 ÷ 9 = 12

Dividing by 10
10 ÷ 10 = 1
20 ÷ 10 = 2
30 ÷ 10 = 3
40 ÷ 10 = 4
50 ÷ 10 = 5
60 ÷ 10 = 6
70 ÷ 10 = 7
80 ÷ 10 = 8
90 ÷ 10 = 9
100 ÷ 10 = 10
110 ÷ 10 = 11
120 ÷ 10 = 12

Dividing by 11
11 ÷ 11 = 1
22 ÷ 11 = 2
33 ÷ 11 = 3
44 ÷ 11 = 4
55 ÷ 11 = 5
66 ÷ 11 = 6
77 ÷ 11 = 7
88 ÷ 11 = 8
99 ÷ 11 = 9
110 ÷ 11 = 10
121 ÷ 11 = 11
132 ÷ 11 = 12

Dividing by 12
12 ÷ 12 = 1
24 ÷ 12 = 2
36 ÷ 12 = 3
48 ÷ 12 = 4
60 ÷ 12 = 5
72 ÷ 12 = 6
84 ÷ 12 = 7
96 ÷ 12 = 8
108 ÷ 12 = 9
120 ÷ 12 = 10
132 ÷ 12 = 11
144 ÷ 12 = 12

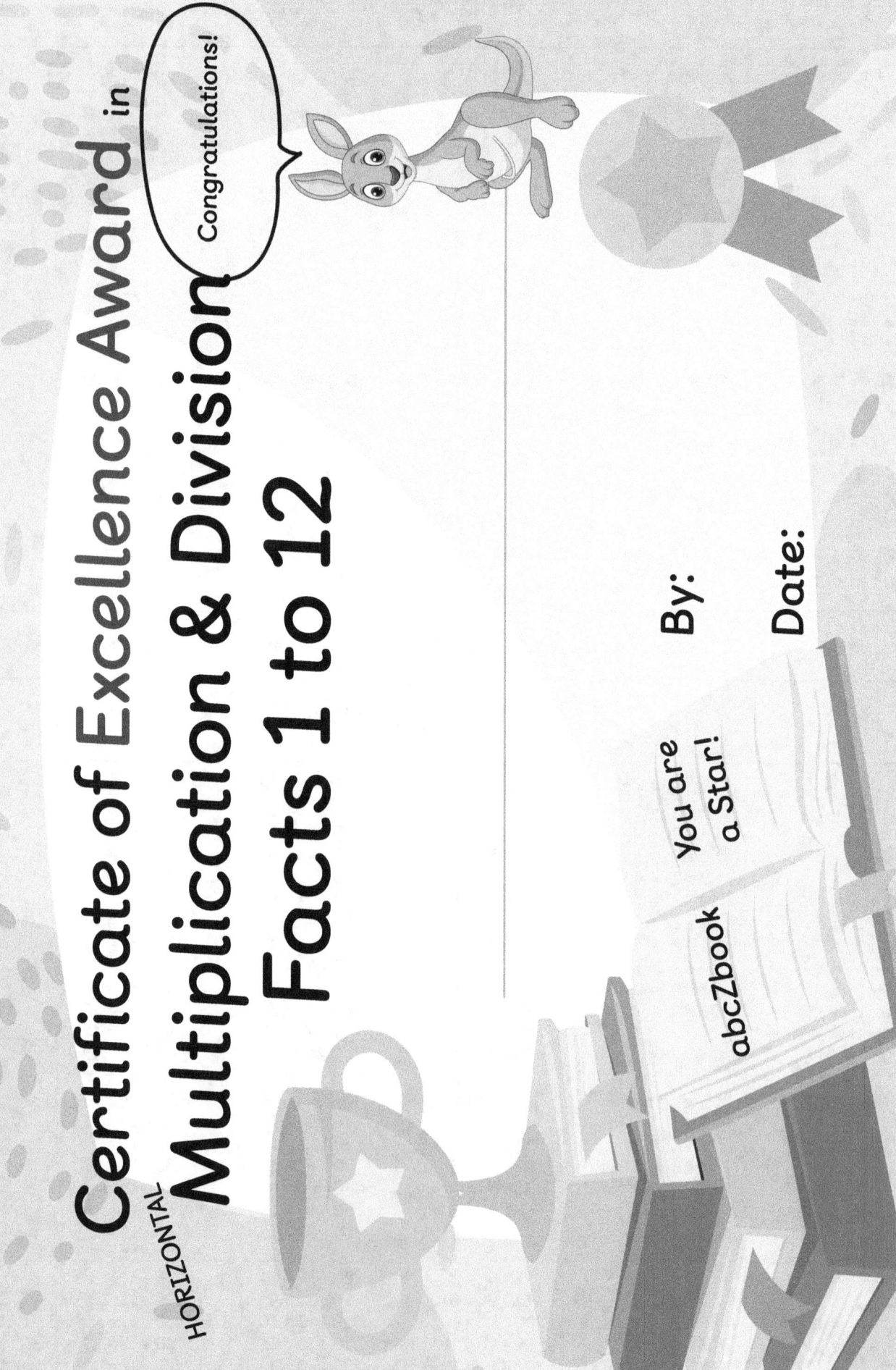